Mary Gribbin and John Gribbin are both graduates of the University of Sussex, where John Gribbin is now a Visiting Fellow in Astronomy. Their earlier co-authored books include *Too Hot to Handle?*, about the greenhouse effect, and *Time and Space*. Mary Gribbin mainly writes science books for young readers, including *Hearing*; John Gribbin mainly writes science books for adults, including *Schrödinger's Kittens*. They have two sons and live in Sussex.

For William Murray with love and thanks

Being Human

Putting people in an evolutionary perspective

MARY & JOHN GRIBBIN

PHŒNIX

A Phoenix paperback
First published in Great Britain by J M Dent Ltd in 1993
This paperback edition published in 1995 by Phoenix,
a division of Orion Books Ltd,
Orion House, 5 Upper St Martin's Lane, London WC2H 9EA

A CIP catalogue record for this book is available from the British Library.

ISBN: 1 85799 378 0

Typeset at Deltatype Ltd, Ellesmere Port, Cheshire

Printed and bound in Great Britain by
The Guernsey Press Co. Ltd, Guernsey, C.I.

CONTENTS

The average human protein is more than 99 percent identical in amino acid sequence to its chimpanzee homolog . . . the nucleic acid sequence difference of human and chimpanzee DNA is about 1.1 percent.

Mary-Claire King and A. C. Wilson
Science, Volume 188, p. 112, 11 April 1975

Life is what happens to you
when you are busy making other plans.

John Lennon,
'Beautiful Boy', 1980

ACKNOWLEDGEMENTS

This book is a reworking of themes developed originally in our two earlier collaborations, *Children of the Ice* and *The One Per Cent Advantage*. The sociobiological aspects of our earlier work owed a great deal to the advice and encouragement of Professor Edward O. Wilson, and both books drew on material used by John Gribbin in related BBC Radio 4 series, produced by Michael Bright. Yoko Ono Lennon gave permission for the lines from John Lennon's 'Beautiful Boy' to be reproduced here, and our own two boys, Jonathan and Ben, helped in the keyboarding and editorial aspects of the new version.

Mary Gribbin
John Gribbin
June 1992

INTRODUCTION

What does it mean to be human? Asked that question, most people respond with thoughts about the things that make us special – achievements in art or science, the trappings of so-called civilization – and mark us out from the other species that inhabit the Earth. The idea that humankind is special is so deeply ingrained that even people whose training ought to have opened their eyes can fall prey to the cosy assumption of human superiority. A little while ago, for example, we offered a news report about a recent investigation of human behaviour to a friend who works on a scientific magazine. The work depended on treating human beings according to exactly the same rules of evolutionary behaviour as red deer or elephant seals – a quite defensible thing to do, in the context (if you want to know more, you will find the story in Chapter Seven). Our friend, who shall remain nameless, responded immediately and instinctively: 'I can't use this,' he said. 'You can't treat people the same way you treat animals.'

It took some time to persuade him just how wrong his remark was. We hope that with this book we shall persuade you (as we eventually persuaded him) that, indeed, the *only* way to get a deep understanding of human behaviour is to apply to humankind exactly those rules that have proved so successful in explaining the way other animals behave.

Being human simply means being one variety of animal on planet Earth. Our similarities to other species, with whom we share a great deal of our genetic inheritance, turn out to be more remarkable than the differences. Impeccable modern chemical techniques show that the difference between the genetic material, the DNA, of a human being and a chimpanzee is just 1 per cent. Small wonder, then, that human beings conform very closely to the patterns of behaviour of other animals.

It is the 1 per cent difference in our DNA that has made

human beings out of African apes; and, to human beings, being human is clearly a good thing. In this book we shall show you what it really means to be human, and how our lives continue to be moulded and influenced by our animal inheritance.

But *why* are we here? How is it that human beings are among the more successful forms of life around on Earth today? There are many answers to these questions, depending on the prejudices of the person you address them to. A cosmologist might tell you about the Big Bang in which our Universe was born, and how the chemical elements of which our bodies are composed were cooked in stars. An evolutionary biologist might talk of natural selection, and survival of the fittest. Both have a point. But one often-neglected factor is the weather – or rather, if we are going to be strictly accurate about these things, the changing climate.

Humankind has emerged during an interval of unusual and rapidly changing climatic conditions which put adaptability and intelligence at a premium, and which pushed one species of African ape out of the woods and onto the path to world domination. But then, what were those apes doing in the trees in the first place? They owed their existence to a far more dramatic environmental change which took place roughly 65 million years ago. Had it not been for the death of the dinosaurs, mammals might have remained inconspicuous creatures, occupying small-animal niches in the ecology and scurrying about in the undergrowth. And what was it that killed the dinosaurs? Almost certainly, bad weather.

Mammals had actually been around for more than a hundred million years before the abrupt (geologically speaking) end to the era of the dinosaurs. The two kinds of animal can hardly be said to have been in conflict. There had been a long period of climactic calm and constancy in which the dinosaurs ruled the roost because they were well adapted to existing conditions, and no new variations on the evolutionary theme got much of a look in.

We certainly cannot make any kind of a case that our ancestors were any more intelligent than their dinosaur contemporaries. The traditional picture of the lumbering, stupid dinosaur with huge body and tiny brain tells only part of the story. Many dinosaurs *were* big and stupid; many more, though, were smaller, agile, and with respectably large brains for their body weight, and it is quite likely that they were warm-blooded. There were dinosaur equivalents of hunters like the tiger and the lion, as well as dinosaur equivalents of grazers like the hippopotamus and the elephant. Dinosaurs were not beaten in the evolutionary stakes by smarter mammals, but were wiped out by environmental catastrophe, a change in the rule of the game.

It is a sign of how successful dinosaurs were, in their own time, that they were actually a younger variation on the theme of life than mammals. Mammal-like reptiles existed before dinosaurs came on the scene, and they were displaced by the dinosaurs. We are descended from the *ancestors* of dinosaurs.

And before that? Like the character in *The Mikado*, we can trace our ancestry 'back to a protoplasmal primordial atomic globule. Consequently [our] family pride is something inconceivable.' In this book, however, we do not wish to go back that far. In order to make our story manageable, we start quite late in the evolution of life on Earth, when our ancestors had already emerged from the sea onto the land. This is not merely a narrative convenience, since life on land has been subject to much more in the way of vicissitudes and changing environmental conditions than life in the sea ever was. That is a major reason why evolution on land has proceeded so swiftly in the past few hundred million years, and why we are here to speculate on our origins.

But before we step back to the proper beginning of our tale, and discuss the broad sweep of evolutionary and environmental events that have combined to make us human, perhaps we should spell out, for the benefit of any doubters, the evidence that we are indeed only 1 per cent human and 99 per cent ape.

CHAPTER ONE

A Rather Unusual African Ape

In 1928, a young American zoologist called Harold Coolidge was in Europe visiting various museums to gather data for what was to become a definitive study and classification of the different types of gorilla. Late one afternoon in the Musée Royale de l'Afrique Centrale at Tervuran, Belgium, he casually picked up from a storage tray a small skull that seemed, at first sight, to belong to a young chimpanzee. To his amazement the bones of the skull were totally fused, set in the permanent adult state, with no room for further growth. The skull was not that of a young chimpanzee, but of an adult with a much smaller head than any recognized chimp had. In adjoining trays the young researcher found four similar skulls, and eagerly took note of their measurements, to be reported to the scientific world at large at some future date. By then it was closing time at the museum, and on the way out Coolidge described his findings to the museum's director, Dr Henri Schouteden. He then went off to continue his gorilla study.

Two weeks later a more senior researcher, Dr Ernst Schwartz, came by the Tervuran museum, and the director mentioned Coolidge's discovery to him. 'In a flash,' Coolidge recalled more than fifty years later, 'Schwartz grabbed a pencil and paper, measured one small skull, wrote up a brief description [and] asked Schouteden to have his brief account printed without delay.'* Schwarz's paper appeared in print in April 1929; Coolidge had been scooped by an unscrupulous rival. But this only fired his enthusiasm to find out more about these strange chimpanzees that lived only in the sweep of jungle

* *The Pygmy Chimpanzee*, edited by Randall Susman, p. xii. Full details of books mentioned are given in the Bibliography.

embraced by the bend of the Congo (now Zaïre) river on its southern side, in what was then the Belgian Congo (now also Zaïre).

A prince among chimps

His gorilla study completed, Coolidge turned his attention to chimpanzees. He carried out a comprehensive survey of museum specimens, revising the classifications with the same care he had taken with the gorillas, and reached the conclusion that this hitherto unrecognized 'pygmy' chimpanzee must be a species in its own right, *Pan paniscus*, not simply a subgrouping of the familiar common chimp, *P. troglodytes*. To many zoologists, this seemed an interesting but not a particularly dramatic discovery. After all, in the 1930s the jungles of the Congo still harboured many species that had not been identified and classified. What was one more type of chimp? But Coolidge was convinced that he had stumbled upon something of major importance to our understanding, not just of chimpanzees, but of ourselves.

Ironically, Coolidge had already met a living pygmy chimpanzee, without realizing it. This was 'Prince Chim', a chimp in the possession of the great American primatologist Robert M. Yerkes. Yerkes bought the chimp from a dealer in New York in August 1923, and was told that the animal came from somewhere in the east of the Belgian Congo. Although Prince Chim died of pneumonia in July 1924, having been under scientific observation for less than a year, he made such a dramatic impression on Yerkes that he became the subject of a book, *Almost Human*. As the title suggests, Prince Chim was really something special. In the book, Yerkes describes the differences between Chim and the common chimpanzees with which he was already familiar. Chim's bold, alert intelligence made him a leader among chimps, and he readily imitated human activities (including, on one occasion, shaking hands with a young student, Harold Coolidge) and learned rapidly from his experiences. By chimpanzee standards, said Yerkes in his book, 'Prince Chim seems to have been an intellectual genius'.

In the 1920s and 1930s – just as today – primatologists were intensely interested in the closest living relatives of our own species, *Homo sapiens*. Charles Darwin, in his classic *The Descent of Man*, had speculated that the common ancestor of man and the hairy apes (the chimpanzee and the gorilla) probably lived in Africa, and it was generally accepted by the scientists that the chimp and gorilla species are indeed our nearest relatives. We all recognize this instinctively, without the need for scientific study: it is what makes the chimpanzee enclosure at the zoo so fascinating. Man, chimp and gorilla, it seemed, must have shared a common ancestor, a 'missing link', not so far back in the evolutionary past. But what did that ancestral ape look like? A particularly intelligent type of chimp with the unusual ability to walk upright seemed to Coolidge to be an even closer relative of ours than the common chimp. When his detailed chimpanzee study was published in 1933, Coolidge presented the case for *P. paniscus* as a species in its own right, and went further with the speculation that the pygmy chimp 'may approach more closely to the common ancestor of chimpanzees and man than does any living chimpanzee',* not so much a missing link as a *living* link with our past.

Such a dramatic claim might be expected to have made waves not just in the world of science, but in the wider world outside. Almost any new idea about human ancestry seems to provoke fierce arguments for and against. But Coolidge's hypothesis was the exception: the idea was presented in the scientific literature, its existence was acknowledged, and then it sank, almost without trace, to be largely ignored for another half-century. When the relationship between man and the pygmy chimp began to make scientific headlines again in the late 1970s and early 1980s, it was thanks to a completely different line of attack on the puzzles of evolution. Instead of looking at animals from the outside, or at the fossil bones of extinct species, and trying to assess relationships on the basis of morphology, since the 1960s a growing band of researchers has been looking at the molecules that make up the tissues of living animals today.

* Quoted recently by, for example, Adrienne Zihlman in Susman's *The Pygmy Chimpanzee*, p. 179.

The techniques of modern molecular biology make it possible to compare directly the proteins in a small sample of blood or tissue from one animal with the equivalent proteins in a sample from a member of another species. It is now even possible to compare directly the DNA molecules, the molecules of life, from one species with those of another. And these techniques provide a direct measure of the relationships between living species. Obviously, if the samples are from two members of the same species, they are essentially identical; samples from two different species differ by a certain degree, and the greater the difference, the more distant relatives the two species must be. In evolutionary terms, the greater the difference, the further back in time their common ancestor.

When these techniques are applied to samples from our own species and the other primates, they show that the chimpanzees and gorillas are, as Darwin guessed, our nearest relatives: such close cousins, in fact, that less than 2 per cent of the genetic material, the DNA, in your body is different from the DNA in a chimp or gorilla chosen at random. The pygmy chimp is indeed a slightly closer relative of ours than even the common chimp (although, of course, the two types of chimp are even more closely related to each other than either is to us), and compared with the pygmy chimp we are only 1 per cent human and 99 per cent ape. And that in turn means that the common ancestor we share with chimps and gorillas walked on the Earth, or swung from the trees, very recently indeed, only about 5 million years ago. These dramatic discoveries have swung the spotlight firmly back on *P. paniscus* in the past few years. But before we bring the story of the pygmy chimp up to date, we should at least outline the new molecular studies and put them in their evolutionary perspective.

Evolution, molecules and man

DNA is the stuff of life. The initials stand for deoxyribonucleic acid, the chemical name for the material that carries the hereditary message from one generation to the next, a blueprint – decoded by living cells – for the construction of the chemical compounds required to keep the new individual alive and to provide for its growth and health. Molecules of DNA are generally very long chains made up of many smaller chemical subunits. The basic structure of each chain, its backbone, is a repeating sequence of alternating phosphate groups and sugars; the name of the sugar unit, deoxyribose, is what gives DNA its name. Every phosphate group is the same as every other one, and each sugar is identical to all the other sugar units in the chain. But the crucial importance of DNA for life is that each of the sugar units in the chain has one of four other types of molecule attached to it, and sticking out from the chain. There are four types of these molecules, called bases, found in DNA; adenine (A), cystosine (C), guanine (G) and thymine (T). It is a string of bases along a DNA molecule that spells out the genetic message.

All the rest is just superstructure. What matters is the string of bases. In the same way, the message in this book is contained in a sequence of letters. It does not matter what kind of paper the pages are made of, or what colour ink the letters are printed in. Indeed, when we were writing the book the letters appeared first on the screens of our word processors, not printed on paper at all. These symbols, or their equivalents, were subsequently processed electronically several times before reaching the pages of the book you now hold, but their message has always reminded the same. So, for our present purposes we can ignore the sugars and phosphates and think of a DNA chain as simply a string of letters in the genetic four-letter alphabet, such as AGCCATGTCATT, and so on. 'And so on', in fact, with a vengeance, for if all the DNA in one normal cell of your body could be extracted and laid out in a line it would stretch a distnce of 175 cm, even though each molecule is less than a millionth of a centimetre thick. If all the DNA in all the cells in your body could be unravelled and laid out end to end, it would stretch for between 10 and 20 million kilometres, but a length of DNA sufficient to reach from the Earth to the Sun would weigh no more than half a gram.

The way DNA is fitted inside the cell is a masterpiece of packaging, a story we can only touch on here. The DNA in one cell is not a single, continuous molecule, but is arranged in several separate chains, called chromosomes, in which the DNA is coiled and twisted back upon itself to make thick, short rods of genetic material. In all human cells there are 46 chromosomes, so the average 'molecule' of human DNA would be 3.8 cm long if an average chromosome rod could be unfolded and stretched out. (In fact, the coiling of DNA into 'supercoils' within the chromosome makes each tiny rod so compact that all 46, laid end to end, would be only 0.2 mm long. Some chromosomes are much bigger than others – in human cells, the largest are 25 times as big as the smallest – so the 'average' used in the calculation is purely a hypothetical example.) Such a length of DNA includes 100 million sugars, each with its own base attached: a message amounting to 50 million letters of the genetic alphabet, since the sugars and their attached bases come in pairs. In all, the 46 human chromosomes contain some 2.5 billion base pairs, a coded message which, if typed out in print the same size as the letters on this page, would fill a million of these pages. Even allowing for the limitations of a four-letter alphabet compared with the 26 letters of the English alphabet, this is still an impressively long message which can carry a great deal of information. The structure of DNA ensures both that this whole message can be faithfully copied and passed on to offspring (new cells or new individuals), and that *parts* of the message can be translated, as required, and the information they contain used in the processes that maintain life.

Take cell reproduction first. So far, we have talked about DNA in terms of a single, long molecule. But of course, DNA usually comes in the form of paired molecules, entwined around each other to form the famous double helix. In the double helix, the two DNA chains are linked by chemical bonds which hold the bases on the two molecules together. But the bases pair up in only two specific ways: A always matches up with T, and C always matches up with G. Every G on the strand is 'mirrored' by a C on the other; every A sees its reflection in the form of a T. The two strands of the double helix are, in a chemical sense, mirror images. And if the double helix is unravelled, each strand can rebuild its partner by collecting the raw materials from the chemical soup inside the cell in such a way that the bridges between the two halves of the new helices are made up solely of A – T and C – G bonds.

This happens every time a living cell copies itself and splits in two, a normal feature of growth. Each DNA molecule untwists, and the separated helices gather on opposite sides of the cell. Each helix builds itself a new partner, and finally the cell divides into two identical daughter cells, each with an exact copy of every one of the original DNA double helices.

What of the role of DNA as the genetic blueprint? Something a little more subtle than wholesale copying is required here. Each chromosome, with its 50-million-letter set of instructions, carries far more information than is likely to be needed by the living cell at any one time. But within each chromosome there are subunits, distinct regions of DNA that have their own 'start' and 'stop' markers, each of which codes for one particular chemical process. Such a stretch of DNA is called a gene, and the message between the start and stop markers is the code that tells the cell how to manufacture one type of protein molecule. Proteins provide both the structure of a body and the 'engineers' that keep it working. Your body is made largely of proteins; the haemoglobin molecules that carry oxygen around in your blood (to take just one example) are protein molecules. When a segment of DNA – a gene – is active, the appropriate stretch of the double helix in one chromosome is untwisted, and the genetic message is read by chemical 'workers' inside the cell and transcribed temporarily into a short stretch of a molecule known as RNA (like DNA, but with a different sugar in its backbone). Finally, the RNA message is used to construct the necessary protein molecule. The RNA is then broken up, and its components are reused in reading off another DNA message. This process is going on all the time in all the cells of your body, keeping everything ticking over just the way it ought to. But there is an even more important, and still only poorly understood, role played by the DNA and its protein engineers.

How did your body get to be the way it is? It started from a single cell, formed when a sperm carrying chromosomes from your father met and fused with an egg carrying chromosomes from your mother. Those 46 chromosomes we mentioned before are actually 23 pairs, and in each pair there is what seems to be a duplication of effort. In one pair of chromosomes, for example, each of the long DNA molecules might carry within itself a gene that determines eye colour. On one chromosome, the instruction might read 'blue eyes'; on the other, 'brown eyes'. In that particular case, the body that carries those genes will have brown eyes. (Like much of this description of the workings of DNA, the assumption that there is 'a gene' for eye colour is a simplification. Probably several genes, perhaps from different chromosomes, contribute to determine eye colour. But the simplified version of the story helps to make our point.) Brown-eye genes are dominant over blue-eye genes. One version of the gene – one *allele* – seems to be wasted. But, when the body that contains those chromosomes manufacturers gametes (egg cells or sperm), new chromosomes are manufactured in a complicated (and poorly understood) process in which the paired chromosomes are cut up and the genetic material from one partner is mixed with that of the other to make two new chromosomes. Just 23 single, new chromosomes then go into each gamete, and when two gametes fuse, the full complement of 23 *pairs* is restored. One member of each pair comes from the mother, and one from the father. But each chromosome in each pair is a mixture of genes that have been cut out of chromosomes inherited from both parents of the person who 'donated' that gamete, and pasted together to form a new combination of genes. Where the gene for blue eyes might previously have sat next to a gene for blond hair, taking another simplistic and hypothetical example, it might now find itself sitting alongside a gene for red hair. And where before it was dominated by a brown-eye gene on the partner chromosome, perhaps in this new body it finds another blue-eye gene (strictly speaking, another blue-eye allele) on the partner chromosome, and with no opposition the character for blue eyes is indeed expressed in the body in which the genes reside. It is the inheritance of genetic material from two distinct parents, plus all this cutting and pasting of genes (usually referred to as recombination), that ensures the enormous variety of individual human beings and of other sexually reproducing individuals. Apart from identical twins (or triplets, or whatever), produced when one fertilized egg divides and copies all its 46 chromosomes, no two human beings who ever walked the Earth have had identical DNA blueprints.

But still, how does a unique fertilized egg – a single cell – develop into a fully functioning human adult, or into the adult form of whatever species? The marvel of the biological development of the adult form from a single cell is the greatest unsolved problem in biology today. It is easy to talk in generalities, and to describe the outlines of what must be going on inside individual cells during development, but we are still very far from understanding how those general principles are put into practice. Nevertheless, the general principles themselves provide crucial insight into how we can be so different from the chimpanzees even though we share 99 per cent of our DNA with them. Clearly, during the growth of a complex body from a single cell, different portions of the DNA message are activated and used at different times in different cells. Something – exactly what we do not know – 'tells' a certain group of cells that they are destined to become, say, the liver. In those cells, only the portions of the DNA message that describe the functioning of liver cells are switched on; although each cell carried the whole 46-chromosome genetic blueprint, it does not 'care' whether or not it inhabits a body with blue eyes, or red hair, or whatever. It functions simply as an efficient liver cell, by operating in accordance with just a few pages from one volume of the 'encyclopedia' represented by all of the human DNA in a cell. How does the liver cell know which pages to read, and to ignore the rest? This still largely unsolved puzzle provides one of the most exciting areas of research in modern molecular biology, for answers to these questions may give us the means to control the cell's mechanism and, for example, prevent the occurrence of cancer.

This skill is not, of course, unique to liver cells. Elsewhere, another group of cells is 'told' to become a leg. The cells grow and extend in one direction to become the embryonic limb, laying down muscle tissue, blood vessels and bone as required. Growth continues after birth until adulthood, and then it stops, presumably in response to another message contained within the DNA blueprint. Other examples will come to mind. What matters from our present perspective is that, clearly, there must be mechanisms within the cells for deciding which parts of the DNA message are to be activated, and when. But there is nothing to provide these instructions except the DNA blueprint itself. So there must be genes that control the behaviour of other genes, turning them on and off, speeding up or slowing down their activity, and generally regulating the processes of growth and development. Since the DNA blueprint is contained in every living cell, those operator genes are also present in every living cell, and can modify its behaviour. And this is the clue to the differences between human beings and the other apes.

Perhaps an analogy will help to make our point clear. Suppose you are baking a cake, using a recipe which reads, in part 'take 200 g of flour and 50 g of butter'. You can take exactly the same words and rearrange them so that the instructions read 'take 200g of butter and 50g of flour'. But the cake you bake according to the second recipe will be very different from the one made by following the original instructions. Or suppose the instructions read 'add 250 ml of water and simmer for 20 minutes'. The outcome would be very different if the instruction were 'simmer for 20 minutes and add 250 ml of water'. Again, these are identical words in a slightly different order. The stretch of DNA that 'tells' a human body 'grow a leg' may be identical to the stretch of DNA that 'tells' a chimp body 'grow a leg'. But if the message is switched on at a different time during development, and for a different *length* of time, the two legs produced will be very different from each other. A small difference in a control gene can produce a big difference in the body that is the end product of development (the phenotype). Again, we are not suggesting that there *is* such a specific, simple instruction in human or chimp DNA, nor that the two equivalent instructions really are identical in the DNA of the two species. But there must be instructions rather like this in our DNA and that of our cousins, and the principle is the same.

Now at last, we can see the significance of the findings from molecular biology that 99 per cent of our genetic material is the same as that of the chimp and gorilla. The molecules which are compared to find the genetic 'distance' between two species include many types of protein, as well as the DNA itself, and always the message is broadly the same. All the proteins in a human body are made from combinations of some of 20 chemical units called amino acids. The haemoglobin molecule, for example, is made of chains of amino acids folded and twisted together in such a way that they provide an oxygen carrier. The order of the amino acids along each protein chain is determined by the gene that describes the construction of that protein. In fact, the whole business of transcribing the DNA message simply translates a chain of bases (AGACGTGTTCA, and so on) into a chain of amino acids. The haemoglobin can function perfectly well if one or two of the amino acids in the chain are swapped for others, and when the proteins from different species are compared this is just the kind of difference that is found. Chimpanzee or gorilla haemoglobin is very, very similar to our own. It carries oxygen just as efficiently. But there are one or two minor, insignificant substitutions in the amino-acid chain, and these must reflect equivalent minor, insignificant substitutions among the bases of the DNA double helix that codes for this protein.

It is the same story for many other proteins. When it comes to the DNA itself, it is not yet possible to map out the entire base sequence of even one human chromosome – all 50 million pairs of letters – and compare it with the equivalent chimp chromosome. But it is possible to carry out a piece of chemical trickery whereby the paired strands of *all* of the genetic material in a sample of human cells are separated and mixed with similarly separated strands of DNA from the cells of chimps or gorillas (or, indeed, other species). The separated strands try to pair up again, but where a strand of human DNA combines with a strand of, say, chimp DNA to form a double helix, the new helix is not held together as strongly as a normal strand from either species. Where the bases on the two strands fail to match up with their proper partners on the other strand (A with T and C with G), there are weaknesses. The strength of the bonding between such hybrid strands of DNA can be measured, and this provides a direct measure of the number of weak links. We do not know exactly which bases are different in the two DNA strands, but we can tell, very accurately, how many bases are different. It is this hybridization technique which shows that there is a 'genetic distance' of about 1.2 per cent between ourselves and the chimpanzees: of every 10 million bases on a hypothetical 'average' human chromosomes, only 120,000 will differ from the bases on the equivalent chimp chromosome. Those small differences ensure, among other things, that the proteins manufactured from the chimp's DNA blueprint are very similar, but not quite identical, to the equivalent proteins manufactured in accordance with the human DNA blueprint.

And if some of those different bases on each of the 46 chromosomes lie in the control genes, which determine development and growth from a single cell, then it is very easy to see how the body that develops is in one case a small, hairy ape and in the other a tall, upright, naked ape.

In the interests of scrupulous accuracy, we should mention two things. First, the chimp actually has 48 chromosomes (24 pairs) to our 46 (23 pairs); molecular studies show that one specific human pair is the same as two chimp pairs fused together. Secondly, we are talking here about differences in the DNA that codes for proteins. There is a lot of DNA in the chromosomes that does not seem to code for anything useful: an intriguing discovery which is the subject of a later chapter. The important point here is that just 1 per cent of the genes that code for proteins in your body is different from the genes that code for protein in the body of a chimpanzee.

How, though, did these differences arise?

A basis for selection

Differences in the DNA of different species, and differences in the DNA of different members of the same species, are the basis of evolution by natural selection. Bodies – *phenotypes* – that are well suited to their environmental surroundings will do well, and will pass on copies of their successful genes to many descendants. Phenotypes that are less well adapted will have fewer offspring, and over the generations the better-fitted genes will dominate. Evolution requires two things. There must indeed be selection of the most suited organisms, with reproduction so that they pass on their successful genes; but that carries with it the implication that there is a variety to select from. And that variety arises from changes in the DNA content of living cells, imperfections in the copying process.

Now, in this book we are not going to address in detail the big questions in evolution, such as the origin of life itself, or the origins of major groups like mammals, or why a human is different from a fish. We are primarily concerned with the way things are on Earth today, and with what the best current ideas in evolution and genetics can tell us about ourselves as human beings. There is a lively debate among the experts today on how major evolutionary changes occur, and whether these changes are gradual or relatively sudden ('sudden', in evolutionary terms, meaning something that happens over the course of only a few million years). The minor differences in DNA that distinguish human beings from the other apes do not come into this debate: they are clearly a set of small-scale adaptations which have occurred simply through the routine copying – and occasional miscopying – of genetic material.

No copying process is perfect, and although the cell mechanisms that copy DNA are pretty good, they too have their imperfections. Mistakes during the everyday process of cell division responsible for growth are of no great significance here, because they have no effect on future generations. But mistakes during the cell division in which recombination produces new chromosomes for the sex cells, leading to the development of new individuals, are clearly important. They can give rise to completely new genes, variations which have never existed before. And those new genes can then be tested up against the rest if and when the chromosome in which they sit becomes expressed in a new body.

Copying errors can come about in several different ways. One letter of the genetic code might be miscopied, a T, say, going into a stretch of DNA where there ought to be a G. But this is a pretty boring and trivial kind of mutation. Much more interesting things happen during the wholesale cutting and re-splicing of chromosomes that goes on during recombination. A chunk of chromosome might get copied and then inserted into its new home the wrong way round, so that the message reads backwards; this is called an inversion. Or a chunk might be cut out altogether, the daughter chromosome being spliced together across the gap; if this happens, the spare piece of DNA might become a new chromosome in its own right. New chromosomes can also be generated when part of an old one is copied twice. Then there are changes like those that have happened in at least one pair of human genes, where two chromosomes that used to be separate have been spliced together by mistake. (As well as this fusing of chromosomes, there are six inversions which distinguish human chromosomes from chimp chromosomes. No other differences are distinguishable without resorting to the DNA hybridization technique.)

Anything you can imagine happening to a chromosome during the process of recombination surely has happened, and will happen again. The genome is much more flexible than it appears at first sight. But most of these mutations will have little effect on the body they inhabit. For one thing, a distorted version of a gene – a rogue allele – will usually be paired up with a normal allele inherited from the body's other parent, and the rogue will never get expressed. Small, subtle changes will have no major influence on the body. The fact that the haemoglobin in the blood of a gorilla has one amino acid out of a chain of more than a hundred such acids different from the haemoglobin in our blood does not make it any less, or any more, efficient at carrying oxygen. Just occasionally, however, the new form of a chromosome will code for the production of proteins which make the body it inhabits more effective. That body will do well, and will pass on its new genotype to its offspring, who will do well in turn, so spreading the new gene throughout the 'gene pool' of succeeding generations of that species.

Why, though, did three families of ape – ourselves, the chimpanzees and the gorillas – begin to follow different evolutionary paths some 5 million years ago? Almost certainly, this kind of speciation is a result of different groups of an original parent species becoming isolated from one another, either geographically or through adopting different life styles. An isolated group on an island is the obvious example: the differences between closely related species of bird on the Galapagos Islands helped Charles Darwin to his theory of evolution. But an equivalent change might occur if, as we shall discuss in detail shortly, a climactic shift caused an extensive forest region to begin to dry out, so that the forest shrank and became surrounded by grassy plains. For a species of tree-dwelling ape in that forest, there would be two possible ways (at least) of adapting to such a change. The individuals, or family groups, still in the middle of the forest would succeed best as tree-climbers and leaf-eaters. But their relations at the forest's edge might need to adapt to a different diet, and would need to be able to move out over the plains. There would be different selection pressures on the two groups, and as a result they would begin to diverge.

How rapid would the process be? Molecular biologists can now measure these evolutionary changes, and set up a kind of evolutionary clock, or calendar, to put them in perspective. Today's closely related species clearly share a common ancestor: the horse and the zebra are an obvious example. If such an ancestral form can be identified in fossil remains, it is sometimes possible to determine reasonably accurately when it was that the two forms began to go their separate ways. Armed with this information, the molecular biologists can go back to their studies of proteins and DNA from the living descendants of those fossils, and compare one species with another. Originally, in the ancestor, all the DNA and proteins were essentially the same. As the two 'new' species have evolved, the molecular timers inside each of them have clocked up a variety of mutational changes that make the two species what they are today. So the molecular studies can determine directly, by measurements in the laboratory, how many changes there have been to a particular protein over a particular span of time, or, indeed, how many changes there have been in the DNA itself in that time. The astonishing discovery that came out of these tests in the 1960s and 1970s is that, although each type of molecule investigated (each protein) changes at its own rate, the same protein (haemoglobin, for example) changes at roughly the same rate throughout generations of many different species.

There is no particular reason why this should be so, but it has been demonstrated by studying the proteins of many species living today, including our own. The number of differences in the protein chains of haemoglobin between one species and another gives a rough indication of how long it is since those two species shared a common ancestor. A similar study of another protein, called cytochrome *c*, will give you the same date read off from a different calendar, and in most cases the date can be read off from several such calendars. For example, in cytochrome *c*, change accumulates at a rate of 1 per cent of the protein chain every 20 million years, while haemoglobin shows a change of 1 per cent every 6 million years, equivalent to a change of one single amino acid every 3.5 million years. For closely related species such as ourselves and the chimps, the date can also be read off from direct comparisons of DNA. Each individual calendar is only a rough guide, but when they all give essentially the same date the molecular biologists know for sure that they are dealing with a fundamental evolutionary truth. And all these techniques indicate a three-way split between man, chimp and gorilla about 5 million years ago, while the DNA hybridization technique, in particular, says that humans and chimpanzees share common ancestry, a short time after the gorilla line split off.

This discovery of a molecular clock has been one of the greatest breakthroughs in evolutionary understanding since the time of Darwin himself, and we have described the work in a little detail because it is so exciting. But, strictly speaking, it is not directly relevant to our main theme. We are concerned here not just with the grand sweep of evolution and the origin of species, but with the way animals, especially human animals, behave today. Clearly, the behaviour of animals depends on the coded DNA message in their genes. We should expect our own behaviour also to depend, at least to some extent, on our genetic inheritance. But we might have expected to find our genome so different from that of 'lesser' animals that we would not gain much insight into our behaviour by studying them. Far from it. Whether or not you accept the timing of the molecular clock which says that *P. paniscus* and *H. sapiens* shared a common ancestor as recently as 4 or 5 million years ago – indeed, *whether or not you accept the theory of evolution at all* – there is no escaping the fact that 99 per cent of our genetic material is *the same* as the genetic material of the pygmy chimpanzee. And it is this fact alone that establishes, beyond any shadow of a doubt, that the best way to understand what it means to be human is to look at what it means to be an animal on Earth today. Our genetic inheritance is *very* similar to that of some species, and broadly related to that of a wide circle of more distant relatives. If we understand how genes work to determine the behaviour of animals in general, then we shall understand 99 per cent of our own inheritance, and we shall be equipped to identify the 1 per cent that makes us special. But before we begin this search for ourselves, we should take one last look at the pygmy chimp, the nearest living thing on Earth today to our own immediate ancestors. In December 1984, writing in the journal *Science*,* Roger Lewin could be found commenting that the outcome of DNA studies which showed man and chimp to have shared a common ancestry for a short time after the gorilla line split off 'is unexpected'. He would not have been so surprised at this molecular evidence, perhaps, had he been familiar with the work of Adrienne Zihlman and her colleagues, first published in 1978, which argued both from the molecular evidence and from physiognomy the case for just such a relationship.

The living link

The fundamental feature that distinguishes our own species physically from our closest relatives is that we walk upright, on two legs. The shift from four locomotive limbs to two would have been easiest to achieve in a small, light animal. The nearesat living creature to a small, lightly built 'four-legged person' is, of course, the chimpanzee, and for this reason alone Vincent Sarich, of the University of California, had by 1968 already singled out the chimp as the 'model' that should be used by anyone trying to reconstruct the appearance of our immediate ancestor. Sarich, one of the pioneers of molecular-clock techniques, was then involved in the studies that showed for the first time that man and chimp had indeed shared a common ancestor as recently as 5 million years ago. Following in Sarich's footsteps, and working with him in the 1970s, Adrienne Zihlman discovered Coolidge's 1933 paper on *P. paniscus*, and carried out a study of the body build and skeleton of the pygmy chimp in comparison with *H. sapiens* and *P. troglodytes*. This work led in 1978 to the publication in the journal *Nature* of a paper with the clearest possible title: 'Pygmy chimpanzees as a possible prototype for the common ancestor of humans, chimpanzees and gorillas'.†

The paper hardly took the scientific world by storm. In 1978, palaeoanthropologists still did not take the molecular clock seriously. But, not long after the paper was published, Donald

* Volume 226, p. 1179.

† The paper was by Zihlman, Sarich and their colleagues J. E. Cronin and D. L. Cramer. It appeared in *Nature*, Volume 275, p. 744.

Johanson and Tim White published their description of a fossil of a previously unknown species which walked in East Africa about 3 million years ago, and which is widely believed to represent our own immediate ancestor. This is the famous 'Lucy', whose story is told by Johanson, with Maitland Edey, in their popular book of the same name. Zihlman and her colleagues claimed that the immediate human ancestor probably looked like a pygmy chimp; now, fossils representing the immediate human ancestor had turned up. The obvious question was, how did those fossil remains compare with the bones of present-day *P. paniscus*? Zihlman has carried out a now-famous reconstruction in which the bones of Lucy and the bones of a pygmy chimp, drawn to the same scale, are juxtaposed alongside each other to give the appearance of a composite creature. The result is very striking. And the conclusion, in the words of Zihlman and Jerry Lowenstein at a symposium held in 1980, was inescapable: 'the earliest known hominids at 3.5 million years may have been only one step away from a small ape like the living *Pan paniscus*'.*

Since 1980, as more data have accumulated and traditional palaeontologists have come to accept the evidence of the molecular clock, the case has strengthened still further. In 1982, a symposium on the pygmy chimpanzee was held in honour of Harold Coolidge, at Atlanta, Georgia. The proceedings were published in book form in 1984.† Unfortunately the pygmy chimp, now the subject of intense scrutiny by anthropologists and evolutionary biologists, is almost on the verge of extinction.

Until recently, the only observations available of the behaviour of pygmy chimps were those of a dozen or so, like Prince Chim, kept in captivity. In the early 1980s, Randall Susman of the State University of New York and a small team of

* *New Interpretations of Ape and Human Ancestry*, edited by Russell Ciochon and Robert Corruccini, p. 691. It is only the skull of the pygmy chimp that is small compared with *P. troglodytes*; the rest of the skeleton is roughly the same size. So the common chimp you see at the zoo will give you a fair idea of the size, although not the detailed bone structure, of both *P. paniscus* and Lucy.

† *The Pygmy Chimpanzee*, edited by Randall Susman, from which the description of the pygmy chimp's characteristics and behaviour in this book is largely taken.

researchers spent 18 months studying *P. paniscus* in the wild, in the Lomako Forest of Zaïre, the pygmy chimp's last surviving natural home. Similar studies have been carried out by Japanese primatologists. Among the remarkable features the pygmy chimps share with *H. sapiens*, but not with the other African apes, are a very long period of sexual receptivity in females, with no clearly defined 'mating season' or short-lived regular periods of sexual activity, and a tendency, at least some of the time, to mate front to front. Mixed groups of males and females are much more important in pygmy chimp society than the all-male bands which are important in the social organization of common chimps. And, of course, there is the pygmy chimp's already well-known ability to walk upright with more ease than his cousin *P. troglodytes*.

Studies of captive pygmy chimps have continued. One of these, Kanzi, seems to be a true modern counterpart to Prince Chim. Kanzi lives at the Yerkes Primate Research Center at Atlanta (named in honour of Prince Chim's one-time owner), where E. S. Savage-Rumbaugh works with him. In *The Pygmy Chimpanzee* she describes how Kanzi frequently combines gestures with his vocalization, pointing at things he is eager to show his teacher; no *P. troglodytes* can do this. He also leads his teachers by the hand to get them to go where he wants, something he was never taught to do and which no common chimp has ever been seen doing. All in all, the behaviour of pygmy chimpanzees, and their response to being trained to communicate with people using symbols, is much more like the behaviour of human infants than the behaviour of comparably trained common chimps. 'Each individual who has worked with both species in our lab', says Savage-Rumbaugh, 'is repeatedly surprised by [the pygmy chimps'] communicative behaviour and their comprehension of complex social contexts that area vastly different from anything seen among *Pan troglodytes*.'*

The case for the pygmy chimpanzee being our closest living relative, and probably retaining many of the features of our common ancestor – a living link with Lucy and our past – is overwhelming. In the words of Roger Lewin, '*Homo sapiens* is

* Randall Susman, *The Pygmy Chimpanzee*, p. 411.

really just a rather unusual African ape.'* We could clearly learn a lot about our past, and probably not a little about ourselves, by studying *P. paniscus* in its natural habitat. But, unfortunately for *P. paniscus*, one of the features that makes us unusual among apes is our ability and apparent inclination to wipe other species from the face of the Earth. The forests in the Lomako area are being cut back to make way for coffee plantations, and the relatively small numbers of surviving pygmy chimps are today hunted both for their meat (considered something of a delicacy) and to provide bones used in local religious practices, as well as to supply the pet trade, zoos and, ironically, scientific laboratories and primate study centres like the one in Atlanta. A live chimpanzee can fetch four or five times as much, on the black market, as a plantation worker earns in a month.

Efforts are being made to educate the local population of the real value of *P. paniscus*, as Kabongo Ka Mubalamata, of the Institut de Recherche Scientifique at Bukavu, Zaïre, told the Atlanta meeting.† In addition, the research institute is urging protection measures, principally the creation of a reserve for the pygmy chimpanzees, and stepping up efforts against the black-market trade and illegal hunting. But these efforts may be too little, too late. There remains a very real risk that the pygmy chimp may become extinct just when it is being recognized as our nearest relative. That would indeed be a sad reflection on human nature. In the rest of this book, we intend to reflect not only on our origins but on human nature in general, and also on how we can use our 1 per cent advantage to overcome our sometimes inappropriate genetic programming. But, having elaborated on our similarities to the other African apes, perhaps the next step should be to highlight the differences: the physical characteristics that do, indeed, make us human.

* Roger Lewin, *Human Evolution*, p. 21.
† Randall Susman, *The Pygmy Chimpanzee*, p. 417.

CHAPTER TWO

Marathon Man

What distinguishes us, physically, from the hairy apes that are such close relatives of ours? Leaving aside the behaviour that makes us 'civilized', the most striking feature of *Homo sapiens* in relation to other mammals (with the exception of the marsupials) is that our babies are the most helpless and immature at birth, and need the longest period of protection while they develop into independent individuals. Childhood is longer for us than for the gorilla or chimpanzee, at both ends of the scale. Our babies are born early, in an unfinished state; if the average body size/duration of pregnancy ratio were the same for humans as for all other primates, then human pregnancy would last for at least a year, perhaps longer, instead of nine months. And, even allowing for this extra period of helplessness, our infants seem to stay infants for longer than their hairy cousins.

Our extended childhood may be related to another striking characteristic of our species. Like the relationship between body size and lifespan, which applies to all other primates. According to this, the average lifespan of a human being 'ought' to be about 30 years; in fact, in the developed world, many people can expect to live well beyond the biblical 'three score years and ten', barring accidents. Our whole lives, not just our infant stages, are stretched out compared with those of our nearest relatives. For comparison, whereas the human gestation period of 266 days is just over 1 per cent of our 70-year lifespan, the gorilla's 257-day pregnancy corresponds to just over 3.5 per cent of its 20-year life, while for a chimp the figures are very similar. Looking at more distant relatives of *H. sapiens*, the lion has a gestation period of 108 days, some 2.5 per cent of its lifespan, while the elephant's impressive 624-day pregnancy (almost 2 years!) corresponds to almost 5 per cent of its life. In terms of total lifespan, as well as time taken for development, humans are the long-distance runners of the primates, designed for staying power rather than sprinting through life. This is more than just an analogy, since it also turns out (as we shall see) that many of the adaptations that stretch the human lifespan also make us well suited to long-distance running. But more of that in due course.

A long period of development and the ability to learn new tricks – an ability extending well into old age – are the key features of *H. sapiens*. So, before we can look at how an understanding of the genetic inheritance we share with those close relatives can help us to understand human patterns of behaviour, we should look in more detail at this long, slow process of development from conception to adulthood. It all goes back to the genetic base mentioned in Chapter One.

The genetic base

Since we are concerned with the development and abilities of just one of the millions of species on planet Earth today, it is as well to be sure just what we mean by the term 'species'. There are several ways of defining a species, but from our present perspective the one that matters is that it is a group of animals that mate with one another and produce fertile offspring. Dogs and bitches mate to produce puppies which grow up to be dogs and bitches; men and women produce offspring which grow up and have children in their turn. Dogs and bitches are members of a single species, and men and women are members of another species. But dogs do not mate with, say, budgerigars: the two are members of different species. And, drawing the line very carefully between close relatives, although a he-ass and a mare can mate to produce offspring, a mule, the mule itself is infertile and can have no offspring of its own. So the horse and the ass are different, albeit closely related species. Any healthy adult male human being could, in principle, mate with any healthy female human being of child-bearing age to produce healthy, fertile children, so every person on Earth, regardless of size, shape or colour, is a member of one species, *H. sapiens*.

There is, of course, a great deal of variation between individual members of a particular species, including our own.

Individuals pass on at least some aspects of their own individuality (size, hair colour, length of index finger, and so on) to their own children. And in the animal world at large the success of individuals, specifically their ability to live long enough to find a mate and reproduce, depends on how well they fit into the environment around them. If some aspect of a particular individual body – a phenotype – helps that body to find food and live to become a parent, then that characteristic has a good chance of being passed on to the next generation. These are the three keys to Darwinian evolution: variation from one individual to another, the inheritance of characteristics passed on from parent to offspring, and the selection of successful characteristics by environmental pressures. When Darwin talked of the 'survival of the fittest', he did not mean the most athletic individuals: he meant, as we hope we have made clear, the ones that fit best into their environment, making good use of the available food, coping with hazards (such as predators and the weather), and finding a mate. The fit of the pieces of a jigsaw puzzle, rather than the image of a super-fit Olympic champion, gives you a better idea of what Darwin was on about.

Very recently (recently, that is, on the timescale of evolution), humans have begun to change the environment to suit themselves instead of adapting, through natural selection, to fit in with the existing environment. This has opened a new chapter in the story of life on Earth, one that began some tens of thousands of years ago when our ancestors began to use tools and make fire. There has not been enough time for this new factor to begin to play a significant part in determining our genetic make-up, although it has, of course, enabled us to spread across the world and increase the total population of human beings on our planet dramatically.

That genetic inheritance, remember, is passed on from generation to generation in the form of 23 pairs of chromosomes, one member of each pair coming from the mother of a new human being and the other from its father. Genes, strung out along the chromosomes, therefore come in pairs. Within each pair an individual gene may be either dominant or recessive with respect to its partner, so that in an individual only one of the genes is necessarily expressed in the phenotype. As a further complication, most of the characteristics of a body as complex as our own are determined by a combination of several genes interacting with one another. Size is an obvious example: children do not grow up to be exactly as tall as either parent. This is partly, of course, because genes provide only a *potential* for growth; the actual size of an adult human being depends also on environmental factors, such as how much food the child gets to eat, working in conjunction with the genetic package the child had inherited from its parents. But more of this later. We would not be so foolish as to claim that genes alone determine even the physical features of an individual, but it is simplest to consider genetic inheritance first and then to look at the interactions between genes and environments. The genetics is complicated enough, with so many characteristics controlled by many genes working together.

Such characteristics are said to be polygenetic: that is, under the influence of two more genes. This is why it is impossible to predict exactly what a new baby will look like, or how it will grow to adulthood, simply by looking at its parents. If the mother has blue eyes and the father has brown, the offspring will not have one eye of each colour, but probably either brown or blue, or perhaps even green, depending on how the recessive and dominant genes match up in its chromosomes. If mother is tall while father is short, the chances are that the offspring will in between; but he or she might very well end up taller than either parent, or shorter than both, even without any environment influence on growth, because of the polygenetic influence of many genes working together to determine the phenotype. Of course, we may say that a boy 'has his father's nose', or that a girl's eyes are 'just like her grandfather's'; even within your own family you can see the inheritability of characteristics. But you cannot predict in advance that one, or any, of your own children will have your nose or their grandmother's hair. Control genes guide our development from conception to maturity, and ensure that we develop as members of *H. sapiens*, not as horses or pigs. Within that broad definition of a species, though, there is plenty of room for manoeuvre.

From conception to birth

The individual human being that we recognize with our senses is a phenotype, the bodily expression of some combination of genes working together. But it is the unique *genotype* that decides how that individual person has been put together, and that new genotype is determined at the moment of conception, when a single sperm carrying a half-set of chromosomes from the father penetrates a single egg containing a half-set of chromosomes from the mother. Once this happens, a complex train of events is set in motion. The surface of the egg immediately changes to prevent penetration by any other sperm, while the nucleus of the sperm, containing its load of genetic material, moves to the centre of the egg and combines with the genetic material it finds there. The egg is now a single cell, called a zygote, carrying a full load of 23 pairs of chromosomes, but less than one-hundredth of a millimetre in diameter.

Just over a day later (about 30 hours after fertilization), the single cell divides to form two cells, and further divisions occur repeatedly over the next few days. All the while, the little ball of cells has been moving down the Fallopian tube towards the uterus. On the sixth day it arrives at the uterus in the form of a hollow ball made up of a few hundred cells, called the *blastocyst*. From now on, different cells have very different roles to play, and develop and differentiate from one another accordingly. As the blastocyst burrows into the lining of the uterus and implants itself, the cells in the outer layer begin a series of changes that will lead to the development of a placenta, drawing nourishment from the mother's bloodstream and passing it on to the developing embryo, and taking waste products from the embryo, through an umbilical cord, and delivering them to the mother's bloodstreamd for excretion or reprocessing. The embryo itself starts off as a small cluster of cells inside the blastocyst. It begins to develop at one end of the hollow sphere of cells, confined within its own inner chamber, a chamber that will become a cavity (the amniotic cavity) filled with fluid in which the developing embryo can grow and be protected from any bumps and bangs from the world outside. (The term 'embryo' is used once implantation has been completed, by the end of the second weeek after fertilization. After 8 weeks (2 months) the embryo has developed to the point where it is given yet another name, the fetus.)

Most of the blastocyst, apart from this initially tiny chamber, is empty. It is, in fact, an evolutionary leftover from the days when our ancestors laid eggs filled with yolk to nourish the developing embryo. The blastocyst 'surrounds' a yolk sac that is no longer there, discarded during the course of evolution. This provides an interesting sidelight on how evolution works. When, through mutations, an improvement on an existing pattern is thrown up, the beneficiaries, the new individuals, still have to carry with them the baggage of their inheritance. Evolution cannot say, 'Aha, I think an animal that brings forth its young alive will be more successful than one that lays eggs', and start from scratch to design a perfect mammal. Instead, a mammal evolves by a series of small changes in the 'blueprint'* that describes an egg-laying reptile. It is, perhaps, a small step for a yolky egg that has already 'learned' to implant itself inside the mother and draw nourishment from her bloodstream to 'learn' to discard the yolk that its ancestors needed. But, judging from the evidence of the development of every human baby, it is a bigger step to change the whole pattern of development of the cells surrounding the yolk to stop them forming a sphere where the yolk used to be. Anyway, the hollow sphere is soon overtaken in size by the developing amniotic cavity and the embryo it contains, and shrivels away to nothing, just as it would if it contained yolk that was being used to feed the fetus.

The mass of cells that will become a new human being starts out as a flat disc. Within a week of conception, the disc begins to differentiate into different kinds of cell – first two distinct layers, and then three. One layer contains the kinds of cell that will develop into the nervous system, hair, skin and nails; on the opposite side of the sandwich is a layer of cells destined to form

* Richard Dawkins, in particular, has expressed reservations about this use of the term 'blueprint' to describe the information passed on to succeeding generations in the genetic code. He says, rightly, that the information is more like that contained in a recipe, 'take the following ingredients and cook them up together', rather than a genuine engineering blueprint which specifies the end product in detail. The environment – the cooking – is also relevant to development. We take his point, but we are happy to live with the more colloquial usage of the term, equally respectably enshrined in the dictionary, meaning a general guide that influences subsequent development.

the internal organs, digestive tract, lungs, and so on. In the middle are the cells that develop to form muscle, the skeleton and the circulatory system. Exactly the same pattern of three cell layers develops at this stage in all animals; it is something we all inherit from a very, very remote ancestor. And, as they must be if a new adult is to be produced and reproduce itself in its turn, the control genes that determine how the cells develop are clearly very much in control right from these earliest stages of development.

At 2 weeks, the now rapidly developing embryo is a mass of cells about 1.5 mm across. This tiny speck of incipient humanity now begins to mould and shape itself into something recognizably human. First, a groove forms across the top of the embryonic disc, and the edges of the groove close up to make a hollow tube. The front of the tube grows, thickening and expanding to form the brain; the back part of the tube will become the spinal cord. Three weeks after conception, the embryo is 2.3 mm long, and is beginning to develop internal organs such as the gut. Another tube forms, and begins to pulsate by about day 24; it will become the heart, pumping 100,000 times or more a day, for as long as the new human being lives. By the end of the first month this simple tube has developed into a four-chambered pump, buds that will become the arms and legs have started to form, and rudimentary eyes, ears, nose and mouth are apparent on the embryo. It is still only a little over 5 mm long, and the mother who carries the embryo may even be unaware that she is pregnant.

Two months after conception, when the mother probably knows that she is pregnant, the embryo has completed all the important stages in laying the foundations for the development of its internal organs and its ultimate outward appearance. From then on the important process is growth, and the development of the individual parts of the body so that they can begin their work when the baby is born; differentiation, the process of creating different kinds of organ and body material out of an initially uniform mass of cells, is largely complete. The fetus, as it is now called, has a skeleton, arms and hands with fingers, legs and feet with toes, internal organs, and a very large head in proportion to its body, complete with the beginnings of eyes, ears, and all the rest. All of this is clearly defined within a miniature person in the making, a fetus abcut 25 mm long and weighing less than a single aspirin tablet, about a gram.

At about this time, just before it qualifies as a fetus, another interesting feature of the embryo reaches its largest size. This is its tail, another reminder of our ancestry. Also during the second month, features resembling the gill slits of a fish appear in the neck region of the embryo, only to develop later into structures of the face and neck. And over a period of several weeks, from the end of the first month into the third month of its life, the embryo/fetus goes through a three-stage process of kidney development. First, it starts to form a pair of kidneys of a type found in primitive fishes and eels, but this soon disappears and a second, more complex version starts to develop. This pair of kidneys also disappears, before the third and final, most complex pair is formed and begins to carry out its function. These and other features of the developing human show that evolution cannot simply discard outmoded designs. A great deal of the DNA in our chromosomes – perhaps the great majority of it – is never expressed at all in the phenotype. Some of this DNA is thought to be 'junk', spacer material or simply rubbish picked up through evolutionary time. But some may well be versions of old genes that used to be essential for our distant ancestors, genes that describe the construction and working of a tail, or gills, or a primitive kidney. Evolution proceeds not by cutting out these old genes and replacing them by new and better ones, but by developing more efficient genes (more efficient in terms of the survival and reproduction of the phenotype) that override and suppress the old material. Although it is not as simple as saying that the developing embryo goes through all the stages of evolution as it develops, as biologists once thought, it is true that the sometimes strange-looking culs-de-sac of development do carry reminders of our evolutionary past.

All such reminders, though, are rapidly fading by the time the embryo becomes a fetus. The rudimentary tail, for example, disappears before birth in 94 per cent of cases, while in the other 6 per cent it is concealed by the flesh at the base of the spine, detectable only as a skeletal appendage in X-ray photographs. During the third month, the fetus begins to move its arms and legs, and it also begins to develop sexual characteristics. As the sex organs form and develop in a male fetus, they release a hormone – *androgen* – which ensures that the body they reside in develops male characteristics. In a female fetus, of course, there is plenty of the female hormone – *oestrogen* – present in the mother's blood, circulating via the placenta and umbilicus, so there is no need for extra female hormone. The baby teeth are being laid down at this time, and the fetus can suck and swallow, and frown and squint. By the end of the third month it is about 23 cm long, from the tip of its head to its buttocks, with its tiny legs folded up in front of it, and it weighs about 15 g.

During the fourth and fifth months the fetus begins to straighten out from its original curled-up position, and the developing organs move down the body cavity into their proper places. After about 4½ months the mother begins to feel the fetus move, and by the end of the sixth month a typical fetus will weigh about a kilogram. By now it measures 30 cm or more, very long and thin for its weight. A layer of fat is only just beginning to develop, and the lungs, for example, are not yet finished; but if a 6-month fetus is born prematurely it does have a chance of surviving, with the aid of modern medical aids such as respirators.

The last three months of pregnancy finish the job off. The fetus fattens up, doubling in size in the last 2 months, and putting on weight at a rate of 30 g a day. At birth, it probably weighs about 3 or 4 kg, although there is a wide range of variation. The lungs are ready to work, and the nervous system is ready to control breathing, heartbeat, sucking and swallowing. On average, the baby will be born 266 days after conception, although in fact only three-quarters of all human births are within 2 weeks of the target date. But although it can breathe, see, hear, suck, cry and wave its limbs about, it is still a very helpless creature, with a head much larger in proportion to its body than that of an adult human being. That large head and the brain it contains is a key reason for the baby being born so soon.

The Peter Pan process

Biology teachers, and their textbooks, often close this stage of the story of human development with a throwaway remark such as 'the genes have done most of their work by the time a baby is born'.* This is a striking way of getting across the important message that the genes responsible for the day-to-day running of your body – making sure there is haemoglobin in the blood, keeping the blood pumping around, producing the right mixture of stomach juices to digest your food, and so on – play a quite different role from the genes responsible for development. It is the development process that produces a distinctively human baby, instead of a member of some other species such as a chimp. As we have seen, DNA is sometimes referred to as the 'blueprint' which specifies a phenotype; extending that analogy, the genes responsible for development from conception to birth are at the same time blueprint, architect and construction engineers. Although some considerable physical development still takes place after birth, the genes that keep the body running during its adult life can then be likened to the janitors, cleaners and maintenance engineers who keep a great building running once it has been constructed. Without those genes performing effectively, the body will die. But without the genes that control development, there would be no living body for them to maintain in the first place.

The crucial differences that make us human, instead of chimpanzee, the 1 per cent advantage we have over our ape cousins, almost certainly lies very largely in those genes that control development. As the example of the tiny difference between our haemoglobin and that of a gorilla shows, the actual running of the end product, the ape or human phenotype, is essentially the same in man, chimp and gorilla. Without doubt, our large brain is a major difference between us and those other apes, and the way this has been linked, in evolution, with the production of babies several weeks before they 'ought' to be born, shows how a small, subtle change in the developmental process can greatly affect the phenotype.

By and large, the bigger an animal, the bigger its brain has to be to control its body. People are pretty large animals – among the primates, only gorillas are larger than us – so we ought to

* For example, see *Human Development & Behaviour* by Jere Brophy and Sherry Willis, p. 39.

have big brains, weighing perhaps just 100 g or so less than the brain of a gorilla. In fact, a typical human brain weighs in at about 1300 g, roughly twice that of a gorilla. And, interestingly in view of the evidence that chimps are our very closest relatives, the chimp, although much smaller than the gorilla, has a brain weighing almost as much as the gorilla's, about 600g. The evolutionary advantages of our large brains are obvious – intelligence, tool-making, speech, philosophy, art and science. We can see that, under the right kind of environmental pressure, increased brain size in relation to body weight will make a species more successful, better 'fitted' to its environment. But this immediately poses a problem for our anthropomorphic image of evolution as a tinkerer who has to make do with the kinds of biological system that are available: there is a limit to how big the brain and head of a fetus can grow in the womb and still be capable of being born without its mother being killed in the process. Birth is difficult enough for human mothers as it is; if their babies were to be carried in the womb for another 3 months or more while they developed to the stage at which other primate infants are born, it would be impossible.

So the evolutionary change which has led to human infants being born early is directly linked to the changes which have increased human brain size so dramatically. At birth, the human brain is one-quarter of its final size, and the skull is still soft, with large 'spots' where the bones have not yet met and fused, to allow for its continued growth. When a rhesus monkey is born, however, its brain is already two-thirds of its ultimate size. One year after birth, the brains of even our nearest relatives, the chimp and gorilla, have reached 70 per cent of their mature weight. A human infant takes 2 years to reach the same landmark, so by this criterion we are born a full year too early! But this central feature of human development at and after birth, the growth of the brain, is only part of the story of how humans fail to grow up like other apes – or perhaps, we should say, how we succeed in hanging on to youth.

Adult human beings share many features with infant apes. Our large brains are housed within a domed skull, with a flat face, very much like those of the baby chimp or gorilla, or indeed like baby kittens and puppies. Later on, though, all these other species develop a distinctive adult shape to the head, with a more or less pronounced snout. We do not. The ratio of the length of our limbs to our body, as well as of the head to the body, is also more like that of an infant ape, or even an ape fetus, than that of an adult. And even our lack of prominently visible body hair, the appearance of the famous 'naked ape', is something that we share with the newborn chimp, or ape fetus, rather than with their adult forms. In very many ways, human development seems to be retarded, so that, leaving aside size, we resemble ape infants rather than ape adults. Like Peter Pan, we never grow up into the adult ape form, although many of us, of course, show traces of further development as we get more hairy with age.

Now, this pattern of characteristics is quite common in evolution, sufficiently so for it to have a special name – *neoteny* – meaning 'holding on to youth'. Only a very small change in the control genes that masterminded development is required for it to occur, which it therefore does very easily. It is a mutation that has happened time and again throughout evolutionary history, a species evolving into a neotenous form, never completing all the stages of development that its ancestors did, then beginning to evolve again along a slightly different path. (Undoubtedly, it has also happened – probably even more often – that a neotenous mutation has occurred but that the resulting phenotype has been no better fitted to the environment than its parents, or than the 'normal' descendants from its parents' generation, and so the new line promptly died out. Variation *and selection* are both crucial to evolution.) Apart from ourselves, the best example of this process is perhaps the Mexican axolotl, a creature rather like a giant tadpole but with four small limbs, that lives and breeds in the water. An ordinary tadpole, of course, cannot breed until it has metamorphosed into a frog. Curiously, though, this sometimes also happens to the axolotl. Under the right circumstances it will metamorphose into an adult form, a salamander, and breed in the usual salamander or frog fashion. The axolotl is a neotenous salamander, an immature form that has 'learned' to live its life and reproduce without developing further.

In the axolotl, neotony is essentially the switching off of the set of genes responsible for metamorphosis. In humans it is less dramatic: we simply develop more slowly, from infancy right through to adulthood and on to the ends of our lives. We are born 'unripe', as it were, take a long time to grow up, and live for much longer than we should. The Peter Pan ape, indeed. This powerful piece of evolutionary magic can be explained quite easily in terms of a change in the workings of a fraction of 1 per cent of our DNA, the small part of our genetic material that controls the rate of development. It is, of course, quite another matter to explain *why* this should have occurred. What were the environmental pressures that made having a larger brain, and all the other features of a neotenous ape, an advantage for our ancestors in eastern Africa some 5 million years ago? Nobody knows for sure why intelligence should then have been placed at an evolutionary premium, although there are two good, front-running theories that are worth a mention. But before we come to them, let us complete the story of the slow development of a human being, from the cradle on, if not quite to the grave.

Early years

By any objective criterion, the newborn human baby is not a pretty sight. Its head has perhaps been squeezed into a distinct point by the process of birth, while its ears have undoubtedly been squashed flat against its skull. It arrives covered in a white protective coating that looks distinctively cheese-like, called vernix, and splotched with its mother's blood. Some are bald; others are covered in a layer of hair that, fortunately, does not persist for long. And yet people respond with tender affection to these nasty little creatures, caring for them and ensuring that they will survive and develop into independent beings. Of course, this reaction is part of our own evolutionary programming. Human beings who did not like the look of newborn babies and left them to die would be very unlikely to produce any offspring of their own that grew up to reproduce and pass on their genes. But mothers who for some reason respond favourably to a newborn infant will tend to be those who have children of their own and pass on whatever it is in their genetic make-up that makes them good mothers. Instincts, including maternal instict, arise, like physical features of the body, through the process of evolution by natural selection. Nothing more is required to explain the basic set of instincts that the newborn human baby is supplied with.

The helplessness of our own offspring at birth is easily seen by comparison with many other mammals. A newborn rhesus monkey can walk, after a fashion, and hold tight to its mother's fur within a few minutes of birth. To take a more domesticated example, a calf can stand and find its mother's udder almost immediately it is born. Even baby kittens, although born blind and helpless, develop independence in a matter of a few weeks, not the years it takes for a human infant to achieve any kind of independence from its parents or other adults.

The basic instincts of the new baby are connected with survival: obtaining food and acting in such a way that parents, and other adults, will respond in a protective fashion. The human infant can suck, cry, hear things, grip things and see things. The cry, thanks to evolution, has a powerful effect on the mother, who cannot bear to leave the infant unattended; very soon the baby also develops a winning smile, to which we are conditioned by evolution to respond. A 'rooting' reflex makes the baby's head turn if its cheek is touched, in the direction of that cheek, an obvious advantage when it is trying to find a nipple to suck on, but something that can cause complications if an inexperienced mother tries to guide an infant's mouth towards her breast by gently pushing the opposite cheek.

The interaction between parents and newborn is, indeed, remarkable enough to merit a more detailed look later, but first we need to outline the rest of the development process. During the first 3 months of its life, the baby develops from a scrawny, helpless creature with an overlarge head into something that has more human proportions, including a respectable covering of fat. It is alert and responsive to its environment, it can hold its head up, and it smiles beguilingly in social situations. It begins to qualify as a human being in its own right, rather than a fetus that emerged from the womb too early. Over the next year or so it undergoes a dramatic physical development, from the early stages of learning to roll over and crawl, to standing while holding onto furniture, standing alone and eventually walking.

There is a great deal of leeway in the ages at which it is 'normal' to achieve these landmarks: some infants, for example, can walk while holding onto furniture at little more than 7 months, while others do not do this until they are 13 months old. But they all go through the same stages in the same order. It is all part of the common genetic inheritance.

The shape of the developing infant also changes. In the womb, the head grew much more rapidly than the body, and at birth the head is still one-quarter of the total length of the infant. In the adult, the head is only one-tenth of the length of the body, because it has grown relatively slowly compared with the rest of the body. This is why a 2-year-old already looks much more 'grown up' than a newborn baby does. The growth of abilities like standing and walking shows how internal processes, especially those connected with the workings of the brain and nervous system, are also developing into an adult state. But physical development, of the body or the nervous system, is, of course, not a particularly notable human activity. All healthy animals can walk about and find food and mate; our cousins the pygmy chimpanzees do all this at a much younger age than we do, and they do a lot of things, such as climbing, rather better than we do. So we shall gloss over most of physical development here, taking it as read that people do learn to stand, walk, run, pick things and carry them, and all the rest. What makes us special, thanks to that 1 per cent difference in our genetic material from that of the pygmy chimp, is our mental development, and that is what we shall concentrate on now.

Development of mental abilities is closely linked with the physical growth and development of the brain. From the quarter of its adult weight that it has at birth, the infant's brain reaches 40 per cent of its ultimate weight by 3 months, half its adult weight at 6 months, and 75 per cent of its final weight shortly after the child is 2 years old. At the same time, the neurones within the brain – the 'wires and switches' of the biological computer – are developing. The neurones themselves get bigger, and the number of nerve connections joining them in a complex network steadily increases. Different regions of the brain develop at different times and at different rates: for example, the parts that control the legs and hands develop markedly between 3 months and 1 year. But, once again, all of this physical development proceeds according to the genetic blueprint, and in the same sequence (in all normal children). The control genes are still firmly in control. At this point, however, we have to tackle a thorny problem, the question of the balance between the role played by the genetic blueprint itself and environmental influences on the developing infant. We assume, of course, that the infant has enough to eat and is well looked after physically; obviously, a starving baby will not develop normally. All of this discussion is about the role of less tangible, but no less real, influences.

This sometimes bitter argument is known as the 'nature/nurture' debate. There are, or have been, scientists who have argued that everything is determined by the genes, and that the environment (nurture) plays little part in development; at the other extreme there have been those who insist that nature (genetic inheritance) counts for very little except physiology, and that the driving force in human development comes from the environment, conditioning the new human being to adopt certain characteristics that will remain important throughout life. As is usually the case with heated debates – and not just scientific ones – the two extremes are unrepresentative. The truth lies somewhere in the grey area in between, and today most students of human development agree that both nature, the genetic blueprint, and nurture, the environment in which the baby develops, are important in determining the kind of human being the baby will grow up to be. There is still debate about the exact balance between nature and nurture, but that need not concern us. What is important is that we take on board the concept of a *duality* of influences, from the genes and from the surroundings, because almost all the misunderstandings about what biologists mean by genetic programming of behaviour arise from the sometimes wilful failure of some people to understand that it is the interplay of the two effects that really matters. The 1 per cent advantage is there at birth – indeed, at conception – but only as a potential; how that potential is expressed depends on the baby's upbringing.

One example, widely quoted by experts on human development, should bring the point home. In the late 1950s the

American psychologist Wayne Dennis visited several orphanages in Iran. In some of these understaffed institutions, although the children were physically cared for and kept clean and fed, there were no resources to provide them with a normal family environment, and the overworked staff carried out their duties like caretakers, providing little human contact.* The babies spent virtually the entire first year of their lives in individual cots, and the sides of the cots were covered by cloths to exclude draughts, so they had nothing stimulating to look at. They were fed with bottles propped up in the cribs, not sitting on someone's knee, and they were picked up only to be washed and bathed, every two days.

Dennis found that the development of these infants was severely retarded. More than half of them could not sit unaided at more than a year old, whereas children reared in normal circumstances do this at 9 months; more than 80 per cent of the 3-year-olds could still not walk alone, something the child in a normal environment almost invariably does before the age of 2; and there were other, similar signs of backward physical development. It looked as if the efficient development of the brains and nervous systems of the babies – all-important in developing these 'motor skills' – had been held back by the dullness of their lives.

So Dennis and a colleague set out to see whether the backwardness could be overcome by providing more stimulation. Thirty orphans, each 1 year old, from the Iranian institutions were tested to measure the stage reached in their development, and then divided into two equal groups. One group was left in the same orphanage routine as before. The babies in the other group were taken out of their cribs for an hour each day and propped up in a playroom with toys to play with. Within a month, the second group showed a marked improvement in motor development over the first.

This is a classic example of the way in which stimulation from surroundings combines with the basic programming laid down in the human genes to produce the complete person. There are many recorded cases in which children have simply

* These were conditions very like those found in 1989 to have existed inside orphanages in Romania.

stopped growing because they received no affection from their parents, for example if one parent deserted and the child was neglected, or if the home was an unhappy one. Time and again doctors have found nothing wrong with such children except for their lack of physical development, and time and again psychologists and doctors have watched in amazement as a change of environment (such as being adopted by loving foster parents) resulted in such a backward child suddenly putting on a growth spurt and catching up with its peers. So, from now on, whenever we talk of the way in which an infant develops, remember two things. First, that although all normal children develop through the same stages in the same order, they do not all do so at the same rate: a wide variety of growth rates, both mental and physical, is covered by the term 'normal'. Secondly, these processes of development can at the very least be hastened or held back, depending on the child's home environment. With that in mind, we can make proper use of the insights into the development of knowledge and understanding in the infant – *cognition* – provided by the pioneering work of the Swiss researcher Jean Piaget.

Towards adolescence

Piaget, who lived from 1896 to 1980, was the most influential developmental psychologist of modern times (although in fact he trained in biology, and never obtained a psychology degree). Starting from observations of his own children as they grew up, Piaget developed the idea of four crucial stages in cognitive development. These were later broken down into more subtle substages, but such details need not worry us here. What matters is the basic idea that a child's capability to understand and interpret its surroundings develops in distinct stages, moving from one stage to the next only when the basic mechanisms of the brain have themselves developed to cope with the new demands placed upon it. In the first, 'sensorimotor' stage, from birth to 2 years, objects exist for the infant only when he or she has direct contact with them. A rattle that is dropped to the floor ceases to exist, as far as the infant mind is concerned; a person is part of a baby's world only when the baby can see, hear, touch or otherwise interact with them. In this stage 'out of sight' is literally 'out of mind'. Then, at about 2 years of age (the exact timing, remember, can vary considerably from one individual to another), there is a quite sudden change to a new stage in which the infant's perception is drastically different. The abruptness of the change, and its qualitative nature, are features of all Piagetian stages. In the second, 'pre-operational' stage, which lasts very roughly until about the age of 7, a child can use language and can understand that objects exist when they are out of sight. A 2- or 3-year-old, for example, will look for a toy that has rolled out of sight behind a chair. But to the pre-operational mind the world at large is still of very secondary importance compared with itself. The world revolves around the infant, so that children in this stage cannot see things from another person's point of view; even if asked to draw a scene as if viewed from a certain perspective they will invariably draw it as they see it themselves. At the same time, the developing mind in the pre-operational stage cannot reason with adult logic. Some children in this stage, for example, learn that living things move, and promptly decide that *anything* which moves must be alive, including a cloud or a car.

Once again, the transition to the next stage is sudden and clear cut – we say that it occurs *roughly* at the age of seven because the actual age varies from child to child. In each individual child, however, there is a very clear transition to the 'concrete-operational' stage, when at last the child develops the ability to use logic in a more adult fashion, when he or she learns about the permanence of properties like number and mass, begins to be able to appreciate another person's point of view, and fully grasps the significance of relatives sizes (A is bigger than B but smaller than C, so C is bigger than B, and so on).*

* This is not to say that children in the pre-operational period are stupid. We recall an occasion when our older son was in this stage, and we visited a student deeply immersed in her work on developmental psychology. It happened that, following a party the night before, there were several bottles on the table. Eager to test her newly acquired knowledge of Piaget's theory, she lined some up in front of our son. Two half-pint bottles stood alongside a pint one. 'Now,' she said, 'if I fill these two liter bottles up with water and the big bottle up with water, will there be more in the big bottle or in both little ones together?' Unhesitatingly, the infant

Finally, at some time after the age of 12 or so, the child passes into the 'formal-operational' stage, and begins to put aside childish things and think like an adult. Abstract reasoning and logic become possible, and he or she develops the ability to investigate systematically all aspects of a problem, eliminating unwanted or incorrect solutions to arrive at the required answer. It is only at this stage that the young human being can deal with hypothetical possibilities, and reason about 'might-have-beens'. So it is only in the formal-operational stage that mathematics, poetry, science and art become part of human life.

It is no coincidence that our educational systems, both in the developed world and in pre-industrial societies, are fitted to these stages of development. You cannot teach abstract mathematics or philosophy to a 7-year-old, so there is no point in trying. (Education is, of course, an important feature of all human cultures. What education can achieve depends on the stage of development a child has reached, and this follows the rules laid down in its genetic inheritance. Let nobody ever tell you that 'genetics' has nothing to do with 'culture'!) Many generations of experience established what could sensibly be achieved, long before Piaget came on the scene. What he did was to show how, in hierarchical fashion, these abilities build one upon the other with each new stage using the groundwork provided by the stage(s) before. And these stages in the development of the mind are related to, and conditioned by, stages in the physical development of the brain itself. Unless the 'wiring' is right, and ready, concrete-operational behaviour, say, is impossible.

Language is such an important feature of being human that we shall deal with it in more detail later. But here it is appropriate to mention the work of the American linguist Noam Chomsky, who revolutionized our understanding of the way children learn to speak with his idea that the human mind is in some way 'pre-wired' for language, and that we are born

reached past her to a quart bottle looming large, to him, in the background. 'That one', he said. He had no idea what the relative merits of two small bottles were compared with one larger bottle, but he could identify the biggest bottle on the table all right.

with a special sensitivity to certain universal features of human grammar. We are programmed to be able to learn to speak, but the actual language we learn (English, Chinese or whatever) depends on the environment in which we are raised. Rare but genuine examples of human beings who grew up in the wild, or were kept in silent captivity, have shown that the ability to learn language persists only for a certain time. If a child has not been given the correct stimulation to learn to speak – which is simply the opportunity to hear other people speaking – by the end of Piaget's pre-operational stage at the very latest, then the child will never acquire language as we use it. Once a human being has acquired language, however, he or she has the ability to learn different languages at a later date. Songbirds, for example, cannot do this. They seem to be programmed with a preference to learn songs from members of their own species, but can be persuaded, if kept apart from adults of their own species when young, to learn other songs instead, including songs no ancestor of theirs has ever learned. Once the pattern is established, though, such birds do not learn new songs when they are adult, even if introduced to members of their own species which sing 'traditional' songs.

This is a very powerful example of the links between nature and nurture, which highlights the single most important ability of human beings: we are able to learn new tricks. The calf, although able to stand and suck its mother's milk within minutes of birth, is so straightjacketed by its genetic programming that it can never learn new tricks. A human being, on the other hand, is programmed not to fill a specific role, but to be able to learn new roles (or new languages) as the need arises. And that is what makes us special. Play is an important feature of the lives of the young of many species, testing out their bodies and practising skills that will later be useful in hunting for food, or in hiding from predators. But humankind takes play-acting seriously. We spend all our life, not just our childhood, playing out roles: at home, at work and in our fantasies. It is all part of our unique ability to adapt and fit ourselves, not to one fixed ecological niche, but to whatever opportunities come to hand. And it begins as soon as we begin to develop into mature adult people, at puberty.

The naked ape

Adolescence is a time of dramatic changes in both physical features and cognitive abilities. Physical changes come chiefly at puberty, the time when, following a signal triggered by a part of the brain called the hypothalamus, the essentially sexless body of a child is transformed, in the space of a couple of years, into a form that is undeniably adult in the sense that it is capable of reproduction. Some of the growth and much of the mental development continue for several years after puberty has, by this strict definition, been completed. The onset of puberty can occur as early as 8 years of age in girls, or may not begin until (rarely in modern society) the age of 17; for most boys it begins later than for most girls, but obviously with such a range of variations some boys complete puberty before some girls of the same age begin the transformation. A reasonable (but arbitrary) definition for the end of adolescence is age 18, at which time most modern Western societies regard the individual as an adult, granting them, for example, the right to vote. Interestingly, the right to reproduce (that is, legal marriage) is often conferred even younger, at 16.

Puberty is associated with a spurt of growth more rapid than anything since early infancy. The average age for peak growth in girls in Europe and North America is 12, while for boys it is 14; boys, however, reach a peak *rate* of growth of 90 mm per year, compared with a typical peak rate for girls of 75 mm per year, so most boys end up taller than most girls. These physical and sexual developments are obvious and familiar to most of us. There is not the space here to go into much detail about the physical changes associated with puberty and adolescence. But most people are far less aware of the equally dramatic mental changes going on at this time, so these we shall mention.

The key cognitive development, as Piaget pointed out, is the ability to think in abstract terms for the first time. It is only in late adolescence that concepts like justice become meaningful to people, and it is only at this stage of development that people become capable of the kind of abstract reasoning involved in most scientific thought, for example. The turmoil caused by the individual's awareness of the dramatic bodily changes he or she is experiencing, and the resulting shift in the individual's self-image, tend to obscure, except to psychologists (and the better kind of teacher), the much deeper changes in reasoning power – in quality as well as quantity – that are happening at the same time. Adolescents begin to try out new things, from different styles of hair and clothing to hero worship of sports or rock stars, or even espousing, usually briefly, extreme religious cults or political movements. This is all part of the testing of their new bodies and new capacity for reasoning. Sometimes the results are naive, in adult terms, but the developing adult learns from experience, another crucial human characteristic.

Inevitably, sex roles become important. In early adolescence boys and girls play in separate, single-sex groups. Later, the two groupings begin to associate with each other, and these interactions lead to the formation of new, mixed-sex cliques. Finally, couples within these cliques begin to associate primarily with one another; the group disintegrates and the important relationship is a one-to-one, boy–girl (or man–woman) interaction.

The result of all this is a sexually mature, physically adult naked ape, equipped with a very large brain, remarkable reasoning power and an ability for complex verbal communication of ideas and information that far outstrips that of any other creature on Earth. Eighteen years of slow growth and development represents the longest childhood of any species we know, but the marathon race of life is only just beginning for an 18-year-old human being. What were the environmental pressures that made this long, slow process of development such a success, in evolutionary terms?

The quality of humanity

Neoteny, the slowing-down of development, is the process which explains *how* our species got to be the naked ape with the big brain. But neoteny does not explain *why* this step in the evolution of our ancestors proved so successful. Obviously it had a great deal to do with the development of our brains. We can see from the fossil record how quickly, and how recently, brain growth took off in our ancestors. We still cannot be sure exactly what pressures placed intelligence at a premium, nor can we do more than guess as to whether our nakedness is a simple side effect of the neotenous development that made for large brains, or whether it carried some selective advantage in itself. But there are two plausible and thought-provoking hypotheses about this phase of human evolution that deserve at least a passing mention, even in a book concerned primarily with the nature of being human today, not with how we got to be the way we are.

The increase in the size of the brains of our ancestors in the direct line that has led to modern people over the past 4 million years is impressive. About 4 Myr,* the varieties of early human whose fragmentary remains survive in parts of Africa had brains with a volume of some 400 cm^3, much the same as the size of the brain of a modern chimpanzee. The probable ancestors of ours from 3 Myr were the australopithecines: *Australopithecus afarensis*, with a brain size only slightly more than this, was probably the ancestor of both the *Homo* line and of the two other australopithecines, *A. africanus*, and the slightly bigger *A. robustus*, which had a brain capacity a little over 500 cm^3. The experts are still debating the exact relationship between these two australopithecines and ourselves, but it was clear that by about 2 Myr, a third species with a still larger brain existed alongside them. Chiefly on the grounds of its brain size, not far off 700 cm^3, this species is the earliest to be give the name *Homo: Homo habilis*.

Within the past 2 million years, through a clear line of evolutionary descent, brain size expanded to an average of some 900 cm^3 in early forms of *H. erectus* (1.6 Myr); 1100 cm^3 in late *H. erectus* (just over 500,000 years ago); no less than 1500 cm^3 in our close relation *H. sapiens neanderthalensis*, which was around until a few tens of thousands of years ago; and 1400 cm^3 on average, in our own species, *H. sapiens sapiens*. For a long time, students of evolution were to some extent over-impressed by this rapid increase in brain size, which they regarded as all-important in determining human nature. They failed to appreciate fully just how much of this increase in brain size was simply due to the increase in body size down this evolutionary line, following the simple rule that bigger bodies

* From here onwards, the abbreviation 'Myr' is used to mean 'millions of years ago'.

generally house bigger brains. *Australopithecus africanus*, for example, had not only a brain the size of a chimp's brain, but a body the size of a chimp's body! A typical *H. habilis*, though, would stand just under 1.5 m tall, and weigh in at under 50 kg. It had a body only slightly bigger than that of a chimpanzee, but a brain almost twice as large as a chimp's. So there was some additional development of the brain, but if simple *size* of brain were indeed all-important, how was it that Neanderthal man, with his bigger brain, was supplanted by ourselves? The question is put in even sharper perspective by the huge variation in size of 'normal' human brains from one individual to the next. Jonathan Swift's impressive 2000 cm^3 makes a puzzling counterpart to Anatole France's mere 1000 cm^3. Humans with very small brains, those suffering from microcephaly, may have a brain only 600 cm^3 in size, scarcely 100 cm^3 greater than a gorilla's. And yet, although of very limited intelligence, they will be very much human beings with the ability, shared by no gorilla, to learn language in a human way.

Modern studies of the evolution of the brain follow two related lines of attack. First, they relate brain size to body size, as far as this is possible. Secondly, thanks to pioneering work by Ralph Holloway, of Columbia University in New York, they are able to analyse the *quality* of the brains that were housed in the fossil skulls they dig up.

Holloway's technique depends on making casts from the skulls and fragments of skulls that survive. In life, the skull fits closely over the brain, so by using the skull as a mould he can make a cast which shows the outlines of the brain that used to live within the skull. Although very faint, these outlines are clearly discernible to the expert eye. From the early 1970s, such casts have thrown up surprises that we can best understand by looking first, briefly, at how the human brain is organized.

Our brains are divided into two almost equal haves, the left brain and the right brain. Each half is made up of four distinct parts, called lobes, and these are identified with specific functions. The frontal lobe, for example, controls movement and is involved in emotions; at the back, the occipital lobe deals with vision; to the side is the temporal lobe, which is important for memory, and on top is the parietal lobe, which seems to play a major part in analysing the flow of information reaching the brain via the senses. For our purposes, the key fact is that in the human brain the parietal and temporal lobes are relatively much larger and dominate the space available inside the skull, whereas in other apes these parts of the brain are much smaller. The main discovery that emerges from Holloway's work* is that the small brains of ancestral hominids that lived at 3 Myr already show this characteristically human structure. In terms of the kind of brain housed within those small skulls, *H. habilis* was certainly already human, and the australopithecines may well have been too.

Now comes the other part of the new insight. Heinz Stephan and colleagues at the Max Planck Institute for Brain Research in Frankfurt made a painstaking study of the size and weight of different parts of the brain from many living species, and compared these with body weights. From all these data they developed a 'progression index' which relates brain size to body weight in closely related species. For an average modern man, the index has a value of 28.8, but this conceals the usual wide variation in human brain size, which makes any value of the index from 19 to 53 'normal'. For a chimp, the equivalent figure is 12.0. The rules developed by Stephan's team have also been used to calculate the equivalent index values for our ancestors. According to these figures, and assuming a weight of a little under 20 kg for *A. afarensis*, our ancestor from 3 Myr already had a progression index of 21.4, as high as some individual human beings today. The figure for *H. erectus* is 26.6. Holloway, using these data, concludes that 'the human brain appeared very much earlier than the time when *H. erectus* emerged', and that '*Australopithecus* and at least one other African primate of the period from 3 million to 1 million years ago had brains that were essentially human in organization ... so far as the subsequent absolute cranial enlargement is concerned, the major mechanism involved, although surely not the only one, appears to have been that as hominids grew larger in body their brains enlarged proportionately.'

So the primary characteristic that makes us human was

* Holloway describes his work in his contribution to the volume of *Scientific American* reprints called *Biological Anthropology*, published by W. H. Freeman in 1975. The quotes below are from that article. His work is put in context by Richard Leakey in *The Making of Mankind* and *Origins* (with Roger Lewin).

already established 3 Myr. This is very soon after the date set by the molecules for the split between our line and the line leading to the chimpanzees, about 4.5 Myr. And the convergence of the evidence clearly points to some key event, or events, that occurred at this time, and set our ancestors on the path that led to ourselves. The kind of development that produced the identifiably human features of the brain are linked with the complexities of human behaviour and social interactions. Evidence of tool-making and the beginnings of 'human' society, in the form of camp sites, are found for the first time at around the same landmark period, about 3 Myr. So, in order to find the environmental pressures that made us human, we have to look for changes that took place around 4.5 Myr which favoured more complex social groupings, the ability for individuals to cooperate with one another, and the ability for them to process complex information from their senses about their surroundings, rapidly and effectively. Once such complex social groups got started, of course, their increasing sophistication would itself feed back to the evolving brains to select those qualities which made for even more effective cooperation. In a similar way, a brain honed to be flexible and absorb new ideas would produce a tool-making culture and would also be refined by the practices of such a culture into an even more recognizably 'human' organ. We can even identify the major environmental changes around 4.5 Myr that must have been instrumental in establishing the proto-human line.

Marathon man – or aquatic woman?

As several critics of the term 'naked ape' have pointed out, we are not in fact hairless: it is just that our hairs are very fine and rather short. But this is something of a red herring, since the absence of long, thick fur is still a striking human characteristic, and the image conjured up by the popular phrase is a good one. Whatever the processes were that made us human, they transformed a tree-climbing ape into an upright walker on the plains. Tree-climbing is actually very good preparation for walking; at least, the kind of tree-climbing practised by our ancestors was. It consists of swinging along underneath branches, hand over hand, and evolutionary adaptations for this kind of lifestyle bring about changes in the arrangement of the body to make it more upright, a flexible shoulder joint, changes in the position of the head and neck on top of the body, and many others.

Life in the trees also sharpened the senses of our ancestors, and put information processing and intelligence, of a sort, high on the list of selective advantages. A tree-climbing ape has to be able to judge accurately which branches will bear its weight, and whether it is safe to jump (or swing) from one branch to another. Accurate three-dimensional vision is essential, and colour vision is also a great asset in helping to select food from the available leaves, fruits and berries. Once adapted to such a lifestyle, it seems unlikely that the successful tree-dwelling apes would ever leave it. But what seems to have happened is that the lifestyle left them.

Around 5 Myr, a major change in the environment of our planet took place. Because of changes in geography caused by continental drift, which blocked off the flow of warm water to polar regions and allowed the poles to freeze, the world entered a long ice epoch within which full ice ages have ebbed and flowed, separated by shorter warm intervals called interglacials. One effect of this, both because water was locked up as ice and because of associated changes in the circulation of the atmosphere, was that the region of East Africa where our ancestors lived dried out. The forests retreated, giving way to more open plains and savannah.

Even so, the most successful tree-apes would still have found a home in the dwindling forests. But their slightly less successful cousins would have been pushed out onto the fringes of the dwindling forests, to make a new home – or die – on the plains. Many must have died; we are descended from the few who first adapted and then evolved to suit the new conditions.

The changes that made our ancestors successful in this new environment included intelligence and adaptability, learning among other things to cooperate with one another and to scavenge the remains of meat killed by hunters such as the large cats. We shall discuss those environmental changes in detail later in this book, but first we give a brief flavour of the kind of debate that surrounds the question of just how environmental changes turned a tree-ape into a human being.

David Carrier, of the University of Michigan, thinks that it was at this point that our relative nakedness also evolved and became important. Although human beings run more slowly than many other animals (our top speed compares with that of a startled chicken), we can keep up a steady jog for hour after hour. This is partly because we are able to keep cool by sweating profusely, but also because we are not covered by thick fur. Most mammals keep cool by panting, which is less efficient as a means to reduce body heat, and although they may be able to sprint very rapidly for a short time, the by-product of this activity is a sharp rise in body temperature, so they soon have to stop to cool off. Carrier suggests that naked skin and sweating evolved alongside running at a steady pace for hours, enabling proto-humans to chase down and kill game, simply by outlasting it. The hypothesis has its attractions, not least since it offers an explanation for an unsual and distinctive human characteristic: the ability to run slowly for long distances, something which, if you think about it, is rather hard to understand in terms of natural selection any other way. But the marathon-man idea is on shakier ground in some other respects, not least because the human body has only a small storage capacity for water, so that a marathon runner runs a great risk of dehydration, especially on the dry African plains.

Which brings us rather neatly to what is surely the most seductive hypothesis for our nakedness, the idea that our ancestors spent a significant part of their evolutionary history as semi-aquatic animals, wading in shallow seas and lakes, and diving for food.

The idea came originally in 1960, from Sir Alister Hardy, and was published in the *New Scientist*.* It was taken up by Elaine Morgan in 1972 in a popular book entitled *The Descent of Woman*, and presented in more complete form by the same author in 1982, in *The Aquatic Ape*. The argument includes evidence such as the proven ability of people to dive efficiently (compared with, say, cattle or lions), the smoothless and hairlessness of our streamlined bodies, the distribution of fat beneath our skin (reminiscent of the blubber of a dolphin, but unlike any other primate) and, shades of the marathon-man

* 17 March 1960, p. 642.

idea, sweating. According to Morgan, however, sweating is a very bad way to control body heat, because it involves the loss of valuable salts through the skin, as well as water. She links sweating with our ability to cry salt tears as an adaptation to eliminate the salts that would otherwise, in a marine environment, build up to excess in the body. And, of course, the 'mystery' of our lack of ability to store water is less of a mystery if we evolved, for a time, within the water itself.

Many of the points raised by Hardy and Morgan merit serious attention, and the aquatic-ape hypothesis is certainly taken more seriously today than when it first appeared. Scientific papers discussing the implications of the aquatic ape now crop up from time to time in respectable scientific journals.* We like the idea enormously, and, accepting that all such ideas about human origins are in a sense 'just so' stories, this is probably the best of the bunch. But it suffers from two serious difficulties, the first of which is to do with timing. The molecular evidence, the brain casts and the environmental changes all point to about 4.5 Myr as the crucial time when the line leading to modern humans became established. Yet by 3 Myr our ancestors were firmly established on the African plains. Was there time for them to have embarked upon an aquatic way of life (which would certainly have been a satisfactory response to the shrinking of the forests and drying-out of the land), only to give it all up and go back to dry land? We simply do not know, but palaeontologists and anthropologists who do take the hypothesis seriously ought to be looking for ways around this difficulty. Morgan herself has not taken full account of the molecular evidence or the brain casts, and gives incorrect dates for the emergence of the human line.

The second problem is why our supposedly successful aquatic ancestors should ever have left the seas at all. There were no major global environmental shifts at the right time to

* There are other examples: A paper called 'The aquatic ape theory: Evidence and a possible scenario', by the Dutch scientist Marc Verhaegen, in *Medical Hypotheses*, Volume 16, p. 17; and a new investigation of the diving reflex by Massaud Mukhtar and John Patrick (*Journal of Physiology*, Volume 370, p. 13). The latter is especially interesting, since their tests show that simply immersing a person's face in a bowl of cold watear enables the average subject to hold his or her breath for nearly 66 seconds, compared with 57 seconds unimmersed.

force the aquatic ape out of the water, although there is evidence that the seaward end of the great African rift-valley system flooded and then dried out between about 6.7 and 5 Myr. The coincidence of this geological dating with the other dates in the story we have outlined is certainly striking, and we believe that more could be made of the connection. The australopithecines, in this picture, would have been the first naked apes, as well as the first ones with human-type brains, forced out of the water and onto the East African plains as the region dried out. But Hardy himself originally suggested that 20 million years of a semi-aquatic existence would be needed to produce the evolutionary changes he and Morgan have latched on to, and there was nothing like this time span available in the light of the modern evidence.

The story has probably been taken as far as it can be by amateurs, no matter how gifted: Morgan herself is a talented writer, the geological evidence comes (hardly surprisingly) from geologists, and Hardy, although a zoologist, never took up the challenge of preparing a complete version of the idea he tossed out on 5 March 1960 to the members of the British Sub-Aqua Club, meeting in Brighton on the south coast of England. One day someone will take it seriously enough to make a thorough professional job of sorting through the evidence and producing a coherent, consistent theory. Until then, we leave you with the best 'just so' story of human origins; perhaps not marathon man but, with a nod to Elaine Morgan, a long-distance (female) swimmer!

Whichever of the 'just so' stories you fancy, however, there is no doubt at all that it was a series of environmental changes that forced our ancestors out of the trees and made us human. And, indeed, the story of the evolution of life on Earth is itself strongly influenced by similar environmental changes, on different timescales. It is time to look back over the broad perspective of evolution, to see just where those tree-apes themselves came from. As we shall explain, we owe our existence, in this perspective, to the death of the dinosaurs; but where did the dinosaurs themselves come from, and why did they die out?

CHAPTER THREE

Death and the Dinosaurs

The story of life on Earth might better be described as the story of death on Earth. Far more species of plant and animal life than those alive today are now extinct, wiped from the face of the Earth because changing environmental conditions made their lifestyles untenable, destroying the ecological niches they inhabited. And yet we are, by definition, descended from survivors. We can trace our ancestry back in an unbroken line to single-celled creatures swimming in the seas of our planet some 3.5 billion years ago. Before that the picture becomes more hazy, but the basic similarity of the life processes in all living things, involving molecules such as DNA, RNA and proteins, suggests that we are all descended from some common ancestor, a collection of molecules that first learned the trick of self-replication.

Replication, of course, is what life is all about. Living things make copies of themselves. But those copies are not necessarily exact replicas of their parents, and that is why life on Earth has been able to evolve and adapt to changing environmental pressures. In each generation, there is a variety of individuals, even within the confines of one particular species. Those individuals that do best – the ones that 'fit' their environment – leave more descendants behind, so that the characteristics that make individuals successful are preferentially passed on to later generations. As long as the environment is essentially unchanging, much of this evolution by natural selection simply makes a species better suited to its ecological niche: monkeys become better adapted to life in the trees, birds develop ever better wings, and so on. But when the environment changes, many superbly adapted individuals, and even whole species, are wiped out because the conditions they are adapted to no longer exist. If all trees disappeared, for example, monkeys would have to learn a new way of life or die. In such circumstances, whether the change is rapid or gradual, the pattern of life is altered. Some species die, descendants of other species come under new selection pressures, and evolve accordingly. Given new opportunities, descendants of the survivors diversify, and new species arise to fill the new ecological niches. From a human perspective, the story of life on Earth is very much one of change and adaptation – a series of changes and adaptations that have led to us – with a great deal of death (of other forms of life) along the way.

Evolution at work

We have already touched briefly on the way evolution works, at the level of the molecule of life, DNA. Some people do still worry about the whole business of Darwinian evolution, however, as questions we have been asked by readers of our earlier books make clear. The main worry that arises in people's minds concerns complexity: how can something as complicated as a human being (or an eye, or the echo-location 'radar' of a bat) have arisen, they ask, 'by chance'? The question is based on a misunderstanding of how evolution works. True, there is an element of chance, in the sense that any small change in the hereditary material – the DNA – that we pass on to our children may be a random alteration of the genetic code. But even this change can only start out from the version of the code the parents carry. If you liken the code to a 'recipe' for, say, a human being, it is equivalent to a fat encyclopaedia written in our familiar alphabet. The random change that is part of the evolutionary process may amount to altering the spelling of a few words in that encyclopaedia, not, in going from one generation to the next, to producing a new, encyclopedia-length recipe, by jumbling up all the letters of the alphabet at random. And, equally important, there is no chance in the vital element of selection that follows, with individuals that fit the environment doing well while those that are poorly suited to their surroundings do badly. It is not as if you took a huge tub of chemicals and stirred them up, waiting for some lucky chance to bring together, in working order, the constituents of a living human being, or even a living cell. Evolution proceeds step by step, each tiny step building on a pre-existing, working system. (Real chance, or luck, does come into the equation for the origin of the first living molecule(s), but there was enough time available, and there were enough molecules around, for life itself to have emerged in this way, genuinely as a result of a lucky accident.) The important point is that evolution has had an almost inconceivably long time to work with: it has been at least 3.5 billion years since life first appeared in the oceans.

Richard Dawkins, in his superb book *The Blind Watchmaker*, puts this in perspective (he also, incidentally, gives the numbers which show how the first living molecules could indeed have appeared by chance). In the span of a few thousand years, at most, people have produced the entire variety of modern breeds of dog from an ancestral wolf species. In each generation, individual people have picked out the dogs they like best (for whatever reason) and used them as breeding stock. This, in a sense, is 'unnatural selection', since it operates in accordance with human whim. But, to a dog, human beings are part of 'the environment', and the process is exactly the same as natural selection. It is a process which has led from wolf to chihuahua in a couple of thousand years. Dawkins suggests that you think of this as a 'distance', equivalent to a single human pace. On that scale, how far would you have to walk along the evolutionary road to go back to the start of evolution on Earth? All the way from London to Baghdad. Evolution has had a *lot* of time to play with. If you are still worried about the details, we recommend Dawkins' book to set your mind at ease. (Indeed, even if you are not worried about evolution, we still recommend *The Blind Watchmaker* as the best non-technical guide to the whole story.) Our aim is not to persuade any doubters that evolution works, but to use the fact of evolution, combined with evidence that the environment of our planet has changed dramatically while life has evolved, to explain how it is that a species of intelligent ape descended from tree-climbing primates should be sitting here wondering about such things, while the dinosaurs, who ruled the Earth for more than 150 million years, have been wiped from the face of the planet.

There are other perspectives. There are still single-celled organisms around, scarcely different from our ancestors of over 3 billion years ago. From their perspective, the story of life on

Earth is one of continuity and stability. Their way of life has not been threatened by any environmental changes, and they are superbly adapted, superbly fitted, to a particular ecological niche. The complexity of multi-celled plant and animal life is simply a curious and incidental by-product of their success.

But of course, that is not our perspective, and single-celled life forms (which, in terms of longevity, are undoubtedly the most successful living forms on 'our' planet) will not get much of a mention in the rest of this book. Indeed, life in the oceans scarcely gets a look in, even though for 3 billion years the story of life on Earth *was* the story of life in the oceans. Plants colonized the land only a little over 400 Myr, rapidly followed by creatures such as millipedes, mites and the first insects. Trees and forests had evolved by about 370 Myr, and the earliest amphibians, our direct ancestors, crawled ashore at about the same time. That is where we choose to begin our story of the changing environmental conditions that made some of the descendants of those amphibian inhabitants of the Palaeozoic era into human beings. But we ought to set the scene by digressing, briefly, to explain how anybody knows what was going on at 370 Myr, or what life forms lived at what times during the subsequent history of the Earth, and how the environment changed.

The record in the rocks

Fossils are the key to understanding the story of life on Earth. Sometimes, when living things (plants or animals) die their remains fall into the ooze of the sea bed, the mud near a lake, or some other place where the pieces can be covered by inorganic material and buried before some passing creature eats them. As more sediments are deposited above the remains, the lower layers are squeezed tight. Eventually, geological forces may mould the sediments into rock, and that rock may carry the imprint of the once-living remains, fossils that are now made of stone but which preserve the shape of the organic remains that fell into the sediments millions of years before. Most organic remains are not preserved in this way, but just a few are.

Such stony copies of living things have been known for

hundreds of years. Leonardo da Vinci, for example, remarked on them in the fifteenth century, and fossils were a cause of great debate and puzzlement among the natural philosophers of the seventeenth century. Ideas developed slowly, but a key step was taken late in the eighteenth century by the English surveyor William Smith, who saw the layers of fossil-bearing strata revealed by new canal-building and coal-mining activity. He realized that different strata contain different, distinctive types of fossil remains, and that rocks of the same age from different parts of Britain (and, as we now know, the rest of the world) can be identified by the characteristic fossils they contain.

In most places, older rocks lie beneath more recently formed rocks, and preserve traces of the fauna and flora of their times. This gave palaeontologists and geologists a relative timescale, and gave Charles Darwin a valuable point of reference as he developed his theory of evolution in the nineteenth century. But it was only in the present century that physicists provided geologists with an accurate clock that could be used to calibrate the geological timescale.

When the geological timescale was first constructed, no dates on it were anything better than guesses. The timescale was divided up, like a geological calendar, into different intervals, but the division was largely on the basis of changes in the fossil record. In some layers of rock there is very little difference between the kinds of fossil found in one layer and those found in the layer above. In other cases, many types of fossil that are present in the lower layer are nowhere to be seen in the upper layer, and 'new' fossils, representing different forms of life, have emerged to take their place (or emerge more gradually as the geologist looks at successively younger layers). The boundaries between geological intervals very often coincide with interesting changes in the pattern of development of life on Earth, precisely because interesting changes in the fossil records are used as the markers for geological intervals.

In the twentieth century these geological intervals have been assigned dates, thanks to the development of radioactive dating techniques. Most things contain traces of radioactive elements, and the strength of the radioactivity of a particular sample of rock (granite, perhaps) decreases as time passes in a regular and now well-understood way. By measuring the radioactivity of a sample of old rock, physicists can calculate how long it has been since that rock was laid down. The figures are not absolutely precise, and slightly different techniques yield slightly different ages, which is why one book may say that the age of the dinosaurs ended at 65 Myr, while another gives the date as 67 Myr, and a third, more cautiously, refuses to give dates but says only that the dinosaurs died out at the end of the Createaceous period (which really is not much help, since one definition of the end of the Cretaceous is 'when the dinosaurs died out'). The dates we use here are largely taken from Steven M. Stanley's book *Extinction*.

Geological time is divided into different slices according to the nature of changes in the fossil record. Big changes – massive extinctions of life – mark the boundaries of more important intervals; little changes – the death of a few species – mark the boundaries of minor subdivisions. Everything that happened before 590 Myr is called the Precambrian, and this seven-eighths or so of the history of our planet is largely a blank as far as our knowledge of life is concerned. Good fossils are only found from later strata (which is why the boundary is set at 590 Myr). The eon we live in, from the Precambrian to the present, is called the Phanerozoic, and this is divided into three eras, the Palaeozoic (from 590 to 248 Myr), the Mesozoic (from 248 to 65 Myr) and the Cenozoic (from 65 Myr onwards). Smaller intervals within eras are called periods, and their names will crop up from time to time in our story. The first period of the Palaeozoic area of the Phanerozoic eon is called the Cambrian (which is why everything earlier is called the Precambrian). Periods are divided into epochs, and epochs into ages, but such subtle distinctions will hardly bother us at all. What we are interested in is why there should have been times, especially at the boundaries between periods, when many forms of life went extinct together. What were the catastrophic changes in the environment which brought doom to the dinosaurs about 65 Myr (thereby opening the way for the rise of the mammals, our own class of animal) and which caused similar extinctions in the more distant geological past?

All things must pass

It is important that fossil hunters do not get too carried away by the appearance of gaps in the fossil record. By its very nature, the process of fossilization provides an imperfect record of the evolution of life, and some of the gaps in that record are there not because species died out, but because their remains happen not to have been preserved. This is important for two reasons. The first is that we have to be sure that many different species were affected before we can claim that there is any geological significance in the disappearance of fossils of a particular type above a certain layer in the rocks. Secondly, there has recently been a fierce debate among evolutionists about the speed at which new species evolve. In some cases, an ancestral form can be identified in rocks separated by millions of years and show little change; then, with no intermediate forms recorded, in the next layer of rocks it has changed into a new species, clearly descended from the parent form but with no intermediate steps visible. Does this mean that the change happened literally in one step, going from one generation to the next? Almost certainly not. The American evolutionary biologist G. Ledyard Stebbins provides a nice example of what may really be happening, in his book *Darwin to DNA, Molecules to Humanity*.

Stebbins imagines that evolution sets to work on an animal the size of a mouse, in such a way that larger animals are more successful – perhaps there is a climatic change which favours larger individuals, because they are better able to retain their body heat. Whatever the reason, suppose that selection acts to favour individuals a tiny bit larger than average, while individuals a tiny bit smaller than average are at a disadvantage. 'Tiny' really is the word that matters, because Stebbins sets the figure so small that it could not be detected in one generation by any human biologist today. It would take 12,000 generations for this hypothetical evolutionary pressure to produce animals the size of an elephant from ancestors the size of a mouse. By assuming that each generation takes five years to reach maturity (longer than the lifetime of a mouse, but less than of an elephant), Stebbins concludes that an elephant-like animal can evolve from a mouse-like ancestor in 60,000 years, while at each step (each generation) the parents and offspring would be indistinguishable by any test we could apply. Mice may indeed be evolving into elephants, or the other way around, before our very eyes, without us noticing.

By human standards, evolution is a slow process. And yet an interval of just 60,000 years is too short to be measured by geological techniques. The emergence of a 'new' life form in less than 100,000 years is, as far as the record in the rocks is concerned, 'instantaneous'. Any fossil remains now being formed in layers near the surface of the ground probably include the bones of many kinds of dog that have lived alongside human beings. Perhaps, in a hundred million years' time, there will be palaeontologists on Earth (visitors from other planets?) who study those remains. To them, it will seem as if the chihuahua appeared overnight, as they move from one layer of rocks to the next. Although they will be able to identify the wolf as an ancestor of the chihuahua, such future palaeontologists will have no hope of reconstructing all the small and subtle steps in the evolutionary processes that converted one animal into the other. Nevertheless, those tiny steps really are the way evolution works.

Important breaks in the geological record are the ones where many different kinds of individual die. Sometimes, life on land is affected more than life in the oceans; sometimes it is the other way round. Sometimes animals suffer while plants survive relatively unscathed, and so on. One way of measuring the severity of an extinction of life forms is in terms of how biologists classify living things. The main division is into five separate kingdoms: animals, plants, fungi and two types of single-celled life form. All known forms of life fall into one of these five categories. No ecological disaster on Earth has yet been severe enough to wipe out a whole kingdom. We can look at the subdivisions of the animal kingdom by following through the lines to which we belong; there are similar subdivisions in other kingdoms.

The next step down from a kingdom is the phylum. From our point of view, the important distinction among animal phyla is between the one for animals with backbones (chordates) and those for animals without backbones (invertebrates). There are five classes of chordates: mammals (to which we belong), birds, amphibians, reptiles and fish. In the past, extinction events have wiped out whole classes, both in the animal kingdom and in others. Such extinctions are rare, and show up clearly as important events in the geological record.

Within the mammal class there are several orders, ours being the primates, each of which subdivides into families. The primate families are the hominids, the great apes, the gibbons, old-world monkeys and new-world monkeys. At the next level comes the genus, in our case *Homo*, and finally the species, *H. sapiens*. In the standard classification, both these last two subdivisions are occupied by ourselves in splendid isolation: an example of human chutzpah, or homo chauvinism, because by any objective standards (perhaps those of the hypothetical extraterrestrial palaeontologist examining our fossil remains a hundred million years from now) it would make more sense to group the two types of chimpanzee (and perhaps the gorilla as well) with *Homo* as one genus (or, if you prefer, to discard the genus *Homo* and classify ourselves as a variety of chimpanzee). But what matters for our present discussion is that species (and genera) often disappear from the fossil record, but leave their cousins alive and well (chimps may very soon go this way, as the tropical rainforest is destroyed, but future fossil hunters will find plenty of bones of *H. sapiens* from strata laid down long after the death of the chimps). The disappearance of a whole family, on the other hand, is more unusual, and if several families disappear at the same time that hints at widespread environmental changes, while the disappearance of whole classes of plants or animals, on land or in the sea, is an even clearer sign of a global catastrophe.

Over the 590 million years since the Precambrian, many families have evolved, gone through changes and faded out, to be replaced by other forms of life. About half a dozen families of marine animals go extinct every million years or so, and such an extinction rate is therefore not unusual in the history of life on earth. But there have been four occasions, at about 438, 253, 213 and 65 Myr, when, many more families (as many as 15 to 20 in a million-year interval) were wiped out. These are the great extinctions which nobody disputes. In addition, there is a marginal case a little before 360 Myr, and many well-established lesser extinctions which affected species and genera but did not wipe out whole families. As you might expect, the four or five most dramatic extinctions each mark a turning point in the evolution of life. The geological record shows that they did indeed occur at times of environmental changes, and in particular it seems that many forms of life have been wiped from the face of the Earth each time our planet has suffered a major cooling and an ice age (or a series of ice ages). We are descended from species that have survived all these crises, and already in that sense we can regard ourselves as children of the ice, although there is a much closer connection between human origins and ice ages, which we shall develop later. The first question that this discovery raises, which has to be answered before we can take the story of life any further, is why the environment should change so dramatically from time to time. Blame today is laid at the doors of two processes. One is slow but certain, and definitely played a part in the environmental changes that have produced us. The other is swift and spectacular; it cannot explain all the extinctions seen in the fossil record, but almost certainly provided a 'last straw' effect that brought the age of the dinosaurs to a dramatic conclusion and ushered in the age of the mammals.

Our changing planet

Continents move about the face of the globe. The geography of our planet is a changing feature of the environment. This discovery was one of the major scientific events of the past fifty years. Although speculation about the remarkable jigsaw-puzzle-like fit between the outlines of South America and Africa goes back at least to the time of Francis Bacon, and the idea of continental drift surfaced in respectable science through the work of Alfred Wegener in the second decade of the twentieth century, it was only in 1962 that Harry Hess, of Princeton University, put forward the key proposal in what became the modern version of continental drift, the theory of plate tectonics.

Before then, supporters of the idea had to argue that the major land masses of the world had reached their present positions by ploughing through the thinner crust of sea floor that lines the ocean basins, rather like big icebergs crunching their way through a thin layer of pack ice. This 'mechanism' simply could not be made to work. The sea-floor crust may be thinner than the crust of material that makes up the continents, but it is still solid rock, and would not conveniently move out of the way as the continents ploughed past. But over the decades more and more evidence had accumulated to show that continents now separated by thousands of miles of ocean had indeed once been part of the same landmass. Fossils formed part of that evidence, similar remains in similar rock strata in many parts of the world. The alignments of the rock strata themselves, and the scars left by the grinding of ancient glaciers across the rocks, showed that, for example, southern South America, southern Africa, south Australia and India had once been locked together, under a blanket of ice. The reason it took so long for continental drift to become respectable was that nobody had a satisfactory mechanism. So Hess's suggestion was welcomed with a sigh of relief in many quarters, and continental drift became respectable almost overnight.

The breakthrough came with the realization that, instead of continents moving through the crust of the ocean floor, it is large 'plates' of mainly oceanic crust that are in motion, the continents simply being carried along on their backs. In some places (notably down the middle of the Atlantic Ocean) there are great cracks in the sea floor, where there are underwater mountains and a seething maelstrom of volcanic activity. New oceanic crust is being forced up out of these cracks in the form of molten rock, which sets and is pushed out on either side, spreading the ocean wider in the process. In other parts of the world (for example, along the western margin of the Pacific), thin oceanic crust is being pushed down underneath the edge of a thicker continent. As the crust is pushed down into the hotter layers of the Earth beneath, it melts; where it scrapes under the continent, mountains are forced upward, and volcanoes belch forth, while earthquakes are common. The islands of Japan are a product of this kind of activity.

Long ago, South America and Africa *were* part of a single supercontinent, now called Pangea, which included almost all of the Earth's continental crust. They were cracked and split apart by the development of the active volcanic feature that is now a spreading ridge in the middle of what has become the

Atlantic Ocean, which is still getting wider. There is no spreading ridge in the middle of the North Pacific, which is steadily shrinking as the American plate is pushed westward. If this process continues undisturbed, the Pacific will eventually shrink into nothing, and North America will collide with Asia.

This is only the briefest caricature of continental drift at work. The theory is now thoroughly well founded, supported by an enormous amount of data from geology. The icing on the cake, to answer any remaining doubting Thomases, came in the 1980s, when range-finding measurements using laser beams from different continents, bounced off artificial satellites orbiting the Earth, made surveying so precise that the drift of the continents could be measured. Europe and America really are getting farther apart, at exactly the rate (a couple of centimetres a year) required by plate-tectonics theory.

Reconstructing the past geography of the globe is a difficult and painstaking task, and it gets harder the further back you look. But geologists are now up to the task, and we can use the pictures they provide without worrying too much about the details of how the reconstructions are made. There is no doubt that about 250 Myr, late in the Permian period, virtually all the continents were grouped into one landmass, called Pangea, which covered the South Pole and stretched in an arc across one side of the Earth almost to the North Pole. Pangea broke up into two continents, Laurasia in the north and Gondwanaland in the south, and these (especially Gondwandaland) fragmented further as the pieces drifted into roughly the places we set them today.

Before Pangea formed, back around 550 Myr when life was not even beginning to move onto land, the world was a very different place. Continents were strung out around the equatorial region of the Earth, and the dominant landmass was an earlier version of Gondwanaland which included large areas of what are now Antarctica, Africa and South America. The poles were covered in ocean, and ice-free. It would have been a warm and pleasant world to live in, except that the continents were all barren deserts, bare rock (and, perhaps, sand), without a trace of soil or life.

The main effect continental drift has in changing the environment of the Earth, making it more or less suitable for life, is that it alters the temperature of the globe. This happens in two principal ways. The direct interrelationship between geography and climate is not entirely straightforward, but in essence it depends on the ease with which warm water from the tropics can circulate to high latitudes. Tropical oceans are warmed strongly by the Sun, which is virtually overhead throughout the year; at high latitudes, the amount of solar heat that can be absorbed depends on the season, and very little of this heat is available in winter. But if warm ocean currents (like today's Gulf Stream) can penetrate to high latitudes, they can help to keep the high latitudes free from the grip of ice (which is why Norway is a more pleasant place to live than Alaska). When there is a continent near the pole (or right over it, as Antartica is today), ocean currents cannot keep the region warm. The land provides a solid base on which snow can settle, building up into great ice sheets. And the shiny white surface of the ice sheets and snow fields reflects away even the summer heat of the Sun, keeping the polar region locked in the grip of ice and lowering the average temperature over the whole globe. By and large, this is bad for life.

The second way in which tectonic activity changes global temperatures is through its influence on volcanic activity. Sometimes there is a lot of tectonic activity, with continents being crushed together and new mountain ranges being thrown up; at other times there is less activity. All of the present-day atmosphere is a product of volcanic activity, gases spewed out over millions of years, in fits and starts. One effect of the atmosphere is that it acts like a blanket around the Earth, holding in warmth that would otherwise escape into space. This is what is called the greenhouse effect, although it works quite differently from the way in which a greenhouse holds in heat. The greenhouse effect is more effective, and the Earth is therefore warmer, when there is more carbon dioxide in the atmosphere (this is now a subject of environmental concern, because human activities, especially the burning of coal and oil, are increasing the atmospheric content of carbon dioxide). Carbon dioxide is one of the chief products of volcanic activity, but it is absorbed both by living things (as part of photosynthesis) and by the oceans, where it is dissolved and may later be laid down as limestone rock. So the concentration of carbon dioxide in the atmosphere has varied over geological time, as the amount of the gas poured out by volcanoes has varied. We have no way yet, unfortunately, of measuring 'fossil' carbon dioxide from the atmosphere of millions of years ago, but for periods for which there is also other geological evidence that the world was particularly warm, we may feel confident that the greenhouse effect was particularly strong at that time.

These two changes, linked with the changing geography of our planet, probably explain many of the climatic changes associated with extinctions of life on Earth. But two other processes are also important enough to mention, before we get on to the meat of our story, the saga of life itself.

Bolts from Heaven

In recent years, the idea that mass extinctions may be caused by bolts from the skies has struck a chord in the popular imagination, and also with many scientists. The battered face of the Moon shows traces of the impact of many solid objects with its surface during its long lifetime. When space probes went into orbit around Mars and sent back detailed pictures of its surface, they showed similar cratering. Venus, hidden beneath a thick blanket of cloud, is more difficult to assess, but there too there is clear evidence of impact cratering. And aerial photography, images from space and careful surveying of the surface of our own planet show that the Earth, too, bears the scars of a bombardment from space. Clearly, the inner part of the Solar System is a dangerous place to be; there are large objects which can collide with planets and moons.

Astronomers know what these objects are. There is a belt of rocky debris, the asteroids, orbiting the Sun between the orbits of Mars and Jupiter. Some asteroids, lumps of rock up to several tens of kilometres across, dive across the Earth's orbit as they move around the Sun; and comets are frequent visitors to the inner Solar System. An impact with any of these would leave a sizable scar on the face of the Earth. Could such a collision create a disturbance sufficient to wipe many families of living things from the face of our planet?

The circumstantial evidence is impressive. Canada provides some particularly good examples, because much of the land surface of Canada is ancient rock with a long history. Furthermore, because of the climate of Canada today, ancient scars are not concealed by a lush covering of forest. At Manicouagan, for example, a ring-shaped feature 70 km wide is interpreted as the scar left by an impact around 210 Myr, when an asteroid more than a kilometre across struck the Earth. New Quebec Crater is a more recent feature, some 5 million years old, a circular lake 3 km wide surrounded by a high mountain rim, looking exactly like a crater on the Moon. This is small compared with some features, such as the circular Deep Bay, 13 km in diameter, whose age has not been measured but which looks like the eroded remains of an ancient impact crater.

There is no doubt that our planet has been struck by bolts from heaven in the past, and that it will be struck again. As recently as 1908 a great expolsion shook Tunguska in the heart of Siberia, devastating forests over an enormous area. Information about the blast leaked out slowly from this wild and unpopulated region, and investigations of the scene were later hampered by the turmoil of war and revolution in which Russia soon became embroiled. But the consensus among scientists today is that the event, equivalent to the explosion of several million tonnes of TNT, or a moderate-sized nuclear weapon, was caused by a relatively small fragment of the icy core of a dead comet (essentially a cosmic iceberg), or perhaps a very small asteroid, entering the Earth's atmosphere and either striking the ground or exploding near the ground as the heat raised by the friction of its passage through the atmosphere vaporized the ice. If the impact had taken place just a little further to the west, Moscow or St Petersburg might have been destroyed, killing the men who were to lead the Russian revolution, and the history of the twentieth century might have taken a very different turn. The Siberian explosion of 1908 did not even leave a crater behind; it must have been modest indeed compared with some of the events that have scarred the Earth.

Very few palaeontologists today would dispute the idea that such impacts have had an effect on life on Earth. Species, perhaps entire genera, could quite easily have been wiped out in the aftermath of such disasters. But whole families? And not just one of two families, but a dozen more? There are fierce arguments among the experts on just how big a role can be assigned to asteroidal and cometary impacts as the cause of mass extinctions. The most widely publicized case concerns the death of the dinosaurs, 65 Myr. At the end of the 1970s, researchers from the University of California at Berkeley found that layers of rock just this age, from several different parts of the world, contain a trace of iridium. Iridium, a heavy metal, is very rare on the surface of the Earth, although it is thought to occur in larger quantities deep inside our planet. It is much more common in the rocky remains of some kinds of meteorite. One explanation put forward for the iridium layer is that it was formed from the debris of an asteroid, a very large meteorite, that struck the Earth just at the time the dinosaurs died out, at the end of the Cretaceous.

In this picture, the debris from the impact rose high into the air, forming a fine layer of dust, laced with iridium, in the upper atmosphere. The dust pall stretched around the world, blocking out the Sun. In the cold and dark beneath the dust, plants and animals died in profusion, and by the time the dust settled out of the atmosphere, forming an iridium-enhanced layer in sediments worldwide, the dinosaurs (and other species), had gone.

It is a chilling, evocative scenario. (Calculations of the so-called 'nuclear winter' that would be likely to follow a full-blooded nuclear war were in fact derived from calculations of this scenario for the death of the dinosaurs.) But there are difficulties with it. First, and most crucially, the dinosaurs did not die out overnight, but over a period of millions of years. Different species disappear from the fossil record at slightly different times, suggesting that the catastrophe that struck them, though rapid by geological standards, was not as quick as the phrase 'death of the dinosaurs' seems to suggest. Secondly, the iridium can be accounted for in other ways. Most notably, it is one of the products of volcanic activity. The famous iridium layer may be telling us, not that a giant meteorite struck the Earth, but that there was an upwelling of volcanic activity at around this time. Volcanic activity can also chill the Earth, by pouring out dust into the stratosphere, or warm the Earth by pouring out carbon dioxide. This would produce a more gradual spread-out change in the environment than a meteorite impact, lasting for hundreds of thousands, or millions, of years. And increased volcanic activity would be a result of changing tectonic conditions, alterations in the drift of continents about the globe.

Another theory, which ought to appeal to everybody but seems to offend proponents of both these ideas, regards meteorite impacts as a possible *cause* of the volcanic activity. Large meteorites could punch holes right through the Earth's crust – especially where the crust is thin, under the oceans – allowing molten rock to flood out, and perhaps initiating a new phase of tectonic activity.

Holes in the sky

The debate continues, and is far from being settled. As if these were not hazards enough for life to contend with, the environment of our planet has changed drastically in other ways, as well. The Earth's magnetic field, for example, is not the fixed and constant guide that mariners would like it to be. Traces of magnetism in old rocks, frozen into the rocks when they were being laid down, show that from time to time the Earth's magnetic field reverses direction completely. Seven hundred thousand years ago, for example, the north magnetic pole was in Antartica. A modern magnetic compass, carried back in time to that epoch, would point south, geographically speaking. In order to get where it is now, the north magnetic pole has swapped places with the south magnetic pole, which used to lie in the Arctic. In such a magnetic reversal, recorded by the fossil magneticism of successive layers of rock, the strength of the field first dies away to zero, and then it builds up in strength, usually in the opposite direction, but sometimes in the same sense as before.

This discovery has caused some confusion among a certain breed of catastrophist. Some popular – but not very scientific – books refer to the Earth toppling over in space as the poles swap positions, or to the crust of the Earth slipping around so that the geographic poles change places. This is *not* what the record of fossil magnetism tells us. It is the dynamo deep inside the Earth, a swirling layer of electrically charged fluid, that changes, presumably because the flow of molten material around the core alters. The geography of the globe stays the same during a magnetic reversal; indeed, we may be experiencing the beginning of such a reversal now, since the Earth's magnetic field is weaker, judging from the record in the rocks, than it was a few thousand years ago. The whole reversal typically takes several thousand years to complete; once the field is established in a certain orientation, it may stay that way for as long as tens of millions of years, or for as little as a hundred thousand years.

Even though the geography of the globe does not change during a reversal, such an event could still be bad news for life on Earth. Our planet is constantly being bombarded by tiny charged particles from space, called cosmic rays. Most of the time, while the magnetic field is strong, these particles are held in two doughnut-shaped regions around the Earth called the Van Allen belts, with some spilling over and being deflected towards the polar regions, funnelling along the magnetic lines of force. There, their major contribution to the environment is to produce the spectacular displays of coloured lights in the sky that we call the aurora. But cosmic radiation of this kind could be quite hazardous to life if there were no magnetic shield to protect us. The radiation is akin to some kinds of radiation produced by nuclear reactors, or bombs, and just as lethal in large enough doses. One of the major medical concerns about the possibility of sending people to Mars, for example, is the worry about how astronauts' bodies might be affected by cosmic rays, especially if the Sun were to produce a large outburst during the several months it would take for a spacecraft to make the journey.

Magnetic reversals are probably not a good thing for life on the surface of the Earth (life in the seas of course, is better protected from cosmic rays). It happens that reversals have been unusually common during the past 85 million years, with the field switching direction nearly two hundred times. There was a similar burst of geomagnetic activity in the Early Devonian, at about 400 Myr, but in between these active intervals reversals were decidedly rare. There was a short burst of magnetic reversal activity near the end of the Cretaceous, with five flips in the course of a million years; perhaps this had something to do with death of the dinosaurs, though if so, it is surprising that nothing so dramatic occurs in the fossil record around 42 Myr, when there were 17 magnetic reversals in the space of 3 million years. Nevertheless, there is some tendency for intervals when the Earth's magnetic field is suffering several flips of this kind to 'coincide' with intervals when an unusually large number of species go extinct. There is also some evidence that magnetic reversals have been more common at times when the Earth has suffered impacts from space. Naturally, this has led some scientists to speculate that the impacts *cause* the magnetic reversals, and that it is a combination of two disasters which then has an adverse effect on life.

This makes a lot of sense. A major impact could very well be just the thing to shake up the currents flowing in the Earth's core, and make them swirl about erratically. It could take a million years or more for the swirling currents to settle down, and this would stretch the period of environmental disturbances caused by an initial short-lived event, the meteorite impact. But it will be very hard to prove that this appealing scenario is what really happened.

Looking further afield, there are other ways in which cosmic events can affect life on Earth. When a star explodes as a supernova, it sends cosmic rays sleeting across space. If a nearby star exploded in this way, then the Earth's magnetic field would provide inadequate protection from the cosmic storm, and many families on land would certainly die. It would be almost as bad if our Sun, which is prone to more modest outbursts known as flares, were to become slightly more active than it is today just at a time when the Earth's magnetic shield was weakened during a reversal. Both these possibilities have been raised by astronomers as explanations of the terminal Cretaceous event. In yet another variation on the theme, some have pointed out that a sudden flood of cosmic rays (caused by a supernova, or by a weakening of the magnetic shield) could disrupt the ozone layer in the stratosphere which shields us from harmful ultraviolet radiation from the Sun. The importance of the ozone layer to life on Earth is familiar to everyone who has followed the saga of chlorofluorocarbons (CFCs), the gases (used in spray cans, refrigerators, and in making the bubbles in foamed plastic) that are implicated in the appearance of holes in the ozone layer over Antartica and the Arctic. Perhaps the dinosaurs died of sunburn and skin cancer.

But all such ideas suffer from the same flaw: they are special pleading, 'one-off' explanations of an event that may be uniquely interesting to us, but which is far from being unique in the long history of our planet. Recently, there have been entertaining attempts to fit this event into a supposed cycle of extinctions.

Cycles of death

Species, genera and families are always going extinct. Four or five families have disappeared every million years or so, throughout the fossil record. But when the numbers of genera going extinct in each million-year interval are plotted on a time chart, intervals with higher-than-average extinction rates stand out. It is always tempting, when confronted with a wiggly line representing some natural, changing phenomenon, to see if there is any regular pattern in the wiggles. Several people have succumbed to this temptation in looking at the pattern of extinctions. Depending on just which peaks in the curve you choose to regard as particularly significant, and exactly which timescale you adhere to (remember that there is some uncertainty in geological dates), there can be found some evidence for peak-level extinctions recurring in waves, at intervals of something between 26 million and 32 million years.

This may be a complete coincidence. A rather nice statistical argument suggests that what appears to be a 26-million-year periodicity should occur at random in the geological record. The argument goes like this. In a system which changes by the same amount at more or less regular intervals, at each change there is a fifty-fifty chance that the system will step up (more extinctions) or down (fewer extinctions). In that kind of sequence, known as a random walk, the most likely interval between *high* peaks is four steps. You can test this, if you are so inclined, with a piece of graph paper, a pencil and a coin. Start from the middle of the left hand side of the paper. Toss the coin, and draw a line which goes one square to the right and either one square up the page or one square down the page depending on whether the result of the toss is a head or tail. Keep going across the page, and you should end up with a series of wiggles about four squares apart.

What has this to do with extinction cycles? Simply that the average geological age lasts for a little over 6 million years. *Small* extinctions, nothing to do with cosmic catastrophes, are spaced about that far apart. So, on the basis of random-walk statistics, unusually high extinction rates ought to occur at roughly 26-million-year intervals (four times six-and-a-bit). The actual geological record over the past 140 million years shows two good peaks that fit the 26-million-year cycle, two less impressive peaks in the right places, and two gaps: much more like random variations than a precise cycle.

This is a shame in a way, since the astronomers had two beautiful ideas to explain a perfect periodicity of between 26 and 32 million years. Both depend on the fact that comets originate in a large cloud of material in the form of a spherical shell around the Sun, far beyond the orbit of Pluto. The ones we see are simply stragglers from this cloud, disturbed and sent shooting in past the Sun by some chance interaction with a neighbouring star. If there were a way to create a big disturbance in the comet cloud, it might rain comets into the inner Solar System for hundreds of thousands of years, many of the objects colliding with the Earth in a burst of destruction far greater than that caused by the impact of a single meteorite. But how do you disturb the comet cloud?

This is where astronomical ingenuity comes in. According to one theory, the cloud is repeatedly and regularly disturbed as the Sun and its family of planets bob up and down (like the needle of a sewing machine passing through a sheet of material) through the star systems that lie in the plane of our Milky Way Galaxy, a great disk-shaped system of stars, dust and gas. According to the second theory, the disturbing influence lies much closer to home. It is either a dark star, in orbit around our Sun and passing through the cloud every 26 million years or so (the so-called 'death star', or Nemesis), or an undiscovered planet, orbiting beyond Pluto and exerting its gravitational influence on the comets (Planet X). Think of a star so far from the Sun that it takes 26 million years to complete one orbit, and you begin to have some idea of how good astronomers are at imagining things.

It would be much easier to accept all this as more than science fictional speculation if there were six clear peaks in the extinction record of the past 140 million years, spaced at precise 26-million year intervals. As you may have guessed, we do not believe it for a minute. But the story of Nemesis and Planet X is important, even if it is science fiction, because it demonstrates how easy it is to think of ways to cause widespread death and destruction among the life forms that inhabit the Earth.

Death is a way of life

From a broad perspective, the death of the dinosaurs at the end of the Cretaceous, and other extinctions which have occurred both before and since, were not so remarkable. Conditions on Earth do change over a long period of time, largely as a result of continental shift. When they change drastically enough there are extinctions which show up in the fossil record. The Late Cretaceous, for example, was a time when the world cooled through these long-term processes, and there was a gradual decline in the number of dinosaur (and other) species over millions of years. Perhaps some bolt from the skies did then have a last-straw effect, ushering in (as we shall see) the age of the mammals.

It is fun to speculate about disaster from above, and to wonder how life can continue at all on Earth, subject to the hazards of cosmic collisions, magnetic reversals, holes in the sky, supernova explosions, dark stars, and so on. Frankly, though, we do not really care – at least for the purposes of the present book – whether any or all of these contributed to the changes at the end of the Cretaceous. It is a fact that the environment changed at that time, and that those environmental (especially climatic) changes had a profound influence on the evolution of life. Without them, we would not be here. Whatever the ultimate cause of those environmental changes, it is the way the changing environment changed the pattern of life that interests us here and now. In later chapters, we shall look in more detail at more recent environmental changes; there, we can talk with much more certainty about why the climate changed. But even then, what really matters to us is not the cause of those climatic changes, but how those changes, whatever their cause, were directly responsible for our own existence.

In telling that tale, we do not need to invoke cosmic catastrophes to explain the broad sweep of environmental changes which are so clearly related to the rise and fall of different forms of life on Earth. Almost always in science, it turns out that the simplest solution to a puzzle is the best, and the simplest explanation of the major extinctions in the fossil record is that they have been brought about by the long, slow processes of continental drift and the resulting environmental changes, especially ice ages. The changing environment has a strong influence on which species (and families) survive and which ones fall by the evolutionary wayside; death is a way of life, in the changing environment of Planet Earth. But life itself, in one form or another, is very persistent. The story of how our ancestors survived, through long intervals of geological time and changing environmental conditions, to become the heirs to the dinosaurs, highlights the tenacity of a family even under adverse conditions.

Our ancestors were the ancestors of the dinosaurs. We, the mammals, inherited the Earth when the dinosaur line died out. It is rather as if a wealthy woman with no children left her estate to the children of her cousin, who shared the same grandparents but had no closer relationship to the benefactor. The analogy is not perfect, because the dinosaurs did not die out in the sense that no dinosaur left any descendants behind. True, many lines did become extinct, but others evolved and adapted to changing circumstances. Birds are descended from dinosaurs; indeed, according to some classifications birds *are* dinosaurs. But birds do not dominate the Earth today. They have achieved their success by occupying specialist niches where they are not in direct competition with mammals: an ironic turnaround from the day of the dinosaur, when mammals survived by occupying specialist niches where they were not in direct competition with the dominant dinosaurs. Who knows, maybe a hundred million years from now the descendants of birds will dominate the Earth and the age of the mammals will seem like a temporary aberration in evolution. Judging from the record of the past few hundred million years, it would not be all that surprising.

Crises in the sea

In the fossil record, one of the clearest examples of the effects of continental drift on life on Earth comes from a time before life emerged from the sea onto the land. At the end of the Ordovician period, around 438 Myr, there were massive extinctions. In terms of the proportion of species affected, this may have been the second-greatest catastrophe to strike during the history of life on Earth to date. Geologists are able to reconstruct maps of the super-continent of Gondwanaland as it was at that time, and they have shown that areas of land that are now South America, Africa (including Arabia), India, Antarctica and Australia formed one continent late in the Ordovician period, and that this continent drifted over the South Pole at that time. These are just the geographical conditions that ought to encourage the spread of ice. Warm water is cut off from the polar regions; land lies ready to form a base on which snow can accumulate and build up into ice sheets; ice sheets, once formed, reflect away incoming solar heat and chill the globe still further.

That is exactly borne out by the geological record. Africa bears the scars of glaciation from the time when it lay over the pole, and the fossils from sediments being laid down at different latitudes also show signs of the cooling. As Gondwanaland moved to cover the pole, species that were adapted to cold-water conditions moved towards the equator, while species that had previously been found in the tropics died out altogether. Cold, spreading outward from the polar regions, explains the mass extinctions at the end of the Ordovician.

The events of 438 Myr certainly provide a neat punctuation mark in our story, since it was during the periods that followed, the Silurian and the Devonian, that life began to move onto the land. Gondwanaland drifted slightly away from the South Pole once more, while pieces of land that now form parts of Europe, North America and Asia were scattered more or less around the equator, and the North Pole was covered by ocean. These scattered pieces of land were edged by extensive shallow seas, where plant life thrived in tidal waters. And it was warm, very probably in large measure because of tectonic activity which spread carbon dioxide around the globe from volcanoes.

Through the greenhouse effect, carbon dioxide helps to warm the world. It is also the basic 'food' that plants need, for photosynthesis. A blade of grass and the trunk of a mighty oak are both made chiefly from carbon, extracted from carbon dioxide in the air. It is hardly surprising that some of those plants in the tidal waters of the shallow seas of the Silurian gradually evolved and adapted to drier conditions, with their descendants eventually spreading onto the land. As the first land plants clustered around the river deltas and tidal flats, natural selection would favour any variation which gave a plant the ability to survive in slightly drier conditions, away from the crowd, where it could get access to unobstructed sunlight, another vital ingredient in photosynthesis.

Animal life followed plant life onto the land for the same reason: competition for space, both physical space and space – a niche – in the ecology. Segmented sea dwellers, ancestors of creatures such as millipedes, made the transition easily and early; cockroaches had evolved and were established well before 300 Myr. Amphibian forms developed from fish, and moved onto the land to eat the plants and insects, not in some natural ascent up the glorious ladder of evolution, but because they found life too tough in the seas. Successful sea dwellers, superbly adapted to their way of life, stayed in the seas. On the margins, less successful life forms found they could eke out an existence by exploiting the newly available resources in the wetlands near the sea; they had to learn new tricks in order to survive at all.

Vertebrates, our direct ancestors, first moved onto the land at the end of the Devonian period, about 360 Myr. This was 'just' after another great catastrophe had struck life in the seas, around 7 million years previously, at the boundary between the Frasnian and Famennian ages of geological time (which is why geologists place a time boundary then). This may be a coincidence, or it may be that the event in the Late Devonian stirred the evolutionary pot and led directly to the emergence of vertebrates from the water.

Towards the end of the Devonian, Gondwanaland once again drifted across the South Pole. Late in the Ordovician, the glaciers had spread over Africa because Africa was directly over the pole; this time what is now South America suffered first because it lay directly over the pole. Geologists see clear evidence of two phases of glaciation, centred in different parts of Gondwanaland at different times. As for the fossils, some rock strata from the Frasnian–Famennian boundary show evidence of a catastrophe that lasted for millions of years, with tropical life forms suffering most. It seems paradoxical, but this is always the case during ice ages. If the whole world cools, from the poles to the equator, then life forms that are adapted to high-latitude conditions can migrate towards the tropics, where the water used to be too warm for them. But life forms that are adapted to tropical conditions have nowhere left to go: there is nowhere for them that used to be warmer than the tropics and has now cooled to be just right for their needs. At least in part, the extinctions in the Late Devonian seem to have been a rerun of the crisis at the end of the Ordovician.

But the plants that were now well established on land do not seem to have suffered greatly at this time. Some of the hardest-hit communities were the inhabitants of shallow seas and estuaries; and the change in fossil remains in some strata occurs so quickly, taking place over a narrow layer corresponding to a very short geological interval, that some palaeontologists, notably Digby McLaren of the Geological Survey of Canada, think that there was an instantaneous, worldwide catastrophe. That bears the hallmarks of the impact of a cosmic object with the Earth. A giant meteorite striking the sea would create tsunamis that devastated the communities of the shallow seas, and vaporize so much water that great swathes of cloud would form, shielding the Sun and cooling the world below.

We know for certain, from the geological evidence, that Gondwanaland moved over the pole and glaciers grew at the time of the Late Devonian extinction; there is circumstantial evidence that the disaster was made worse by an impact event. But whatever the exact causes, what matters is that once again the Earth cooled, and many forms of life went extinct as a result. After the Devonian, the evolutionary action moved to the land, and amphibians appeared. Whether or not the Late

Devonian extinctions gave the amphibians a push, it does provide another neat marker in the story.

Megadynasties and protomammals

The Devonian was followed by the Carboniferous, a period so named because it was a time when great forests spread across swampy, low-lying land; trees that became preserved in the swampy sediments were later compressed by geological forces and turned (eventually) into carbon-rich coal. Some of the coal that we burn in fireplaces, furnaces and factories today is the fossilized remains of trees that lived more than 300 Myr. The Carboniferous period lasted from 360 to 286 Myr; modern geologists often divide the period into two, the Mississippian (from 360 to 320 Myr) and the Pennsylvanian (from 320 to 286 Myr), named after coal deposits of those ages found in two of the present-day states of the USA. The coal locked up in these deposits, and others like them around the world, used to be living trees, and the carbon in those living trees was taken out of the air, by the trees, in the form of carbon dioxide.

The carbon dioxide came from volcanoes, as it does today. The atmosphere does not contain a fixed amount of carbon dioxide which has to be shared out among all the living things that need it; on a geological timescale, the amount of carbon dioxide available varies. The balance between carbon dioxide production as a result of tectonic activity, and the rate at which it is removed from the air by biological processes (in the sea as well as in the land) is one of the imponderables of Earth history. We can guess that the events of the Carboniferous, which led to huge quantities of carbon dioxide being taken out of the air and the carbon being locked away in the rocks, may have made the world cooler than it might otherwise have been since the greenhouse effect would have been weakened. But since we do not know how much carbon dioxide was produced by volcanoes at the time, we have no way of assessing the 'might have been' in the equation. Cooler the globe probably was, as a result of all the coal-making, but cooler than what? We shall never know, but even though our main interest is in the evolution of animal life on land in the 300-odd million years that followed the Carboniferous, the existence of those coal deposits may yet have a part to play in our story.

The tale of animal life in that 300-million-year interval can be summarized in terms of four major developments, what Robert Bakker, of the University of Colorado in Boulder, has called the 'Megadynasties'. In the 1970s, Bakker took palaeontology by the scruff of its neck and gave it a vigorous shake. He overturned long-held views on the nature of dinosaurs and their ancestors, and established beyond reasonable doubt that many dinosaurs were hot-blooded, active creatures. His once-revolutionary ideas are now respectable, although the image of the slow, lumbering dinosaur with a brain the size of a pea persists in many introductions to Earth history. Bakker's book, *The Dinosaur Heresies*, is required reading for anyone curious about the way life has evolved; here, we follow his lead in describing the four Megadynasties.

During the Carboniferous, and the early part of the period that followed (the Permian), the first Megadynasty emerged. Amphibians came first. Like fish, they were cold-blooded; and although they could breathe air and live and feed out of water, they had to return to the water to reproduce, in a life cycle very similar to that of the modern frog or salamander. This did not stop them from reaching respectable sizes: several metres long, with bodies basically resembling that of the alligator, squat legs jutting sideways and bellies close to the ground. Near the end of the Carboniferous, but still within Bakker's first Megadynasty, there was a major evolutionary development: some species developed the hard-shelled egg, freeing them from the need to reproduce in water. The eggs, with their protective cover, could be laid in a nest on land, and the young could develop in a moist world of their own, inside the egg, until they had reached a stage where they could cope with life in the dry conditions outside, and would emerge from the shell. The egg-layers are called reptiles; although there are other distinctions between amphibians and reptiles, this is one that really matters.

These first reptiles, however, were still members of Megadynasty I. They were cold-blooded lumberers, rather like the old image of dinosaurs. Bakker tells how measurements of the spacing between preserved fossil footprints show that these early reptiles were slow walkers, while studies of fossil bones show that the musculature attached to those bones would have made the fastest gait of the greatest predator of the time, *Dimetrodon*, no more than 'a lumbering waddle'. Until halfway through the Permian, the picture stayed much the same. The fossil record shows that there was little diversification of these animals, with just one big family of herbivores and just one big family of predators. This fits the picture of slow-moving creatures with low metabolic rates in a stable world, spreading over the land that had so recently (in terms of geological time) been colonized by plants.

Everything changed, however, halfway through the Permian period, perhaps around 270 Myr, at the time of the Kazanian age. Bakker calls this 'the Kazanian Revolution', a time when 'the entire somnolent world of Megadynasty I passed away'. The reason for the revolution, he argues, is that during the Kazanian some species became warm-blooded. Protomammals had arrived on the scene, and with a higher metabolic rate – including the ability to run fast – they swept their somnolent predecessors from the evolutionary stage.

There is no doubt that things changed dramatically at that time. Where there had been one family of predators, soon there were four; where there had been one family of plant eaters, there were five. Species proliferated, adapting to fill every ecological niche available, and evolution moved faster as a result of competition between species. Measurements of fossil remains confirm that these fast-evolving species were active and must have been warm-blooded. The structure of their bones and the nature of their joints show that the limbs of these creatures were driven by powerful muscles capable of giving them a good turn of speed, while the position of those limbs, straighter and more central under the body than in earlier species, also shows adaptation to a more 'modern' form of locomotion.

Geography also enters into the argument. Animals that live on land are more sensitive to relatively small and short-lived changes in temperature than is life in the sea. The ocean acts as a buffer, taking a long time to warm up, and a long time to cool down, when conditions change. This of course is why an island like Britain, on the edge of the ocean, never experiences either the baking heat of summer or the bitter cold of winter that occurs in a continental interior, in a place like Siberia. In the Kazanian and throughout the late Permian, all the main landmasses of the globe were joined in the supercontinent of Pangea. Gondwanaland was still near the South Pole, but joined to land which stretched continuously up across the equator and almost to the North Pole, with what is now Siberia, as it happens, at the highest northern latitudes. This must have made the whole world cool, even when there was not a full ice age, and warm-blooded animals would have an advantage over their cold-blooded cousins in cool conditions. Cold-blooded animals rely on the warmth of the Sun to stir them into life, powering the biochemical processes that make their muscles work. Warm-blooded creatures, however, derive their heat from inside, using energy from the food they eat, and can stay active and mobile when cold-blooded creatures are sluggish. Even if the first warm-blooded creatures gained only a little warmth in this way, that would still have given them a huge advantage over animals that were totally dependent on external warmth.

The protomammals suffered at least two important extinctions during the remainder of the Permian, and each time the survivors recovered and new species evolved to fill the ecological niches left vacant. Almost certainly, these extinctions were related to the unusual geography of the globe at that time, which must have made it easy for waves of cold to sweep over the land, even if the seas were not severely affected by a short-lived dip in climate. Just possibly, one or both of these extinctions were caused by a meteor impact. But Megadynasty II seemed to have proved its adaptability and warm-blooded worth over a period of more than 20 million years, and to be all set for continued world domination, when the greatest disaster in the geological record struck the Earth. We can best put this, and subsequent events, in a proper perspective by backtracking a little to look at the nature of the protomammals whose promising evolutionary careers were so abruptly cut short, and whose descendants found, in the Triassic period, that they had hazards other than the weather to cope with.

Mammals with a lisp

The warm-blooded, semi-reptilian protomammals that lived in the Permian, some of which survived into the Triassic, were the direct ancestors of every mammal that ever lived. You and us, cats and dogs, bats, whales, mongooses and giraffes, are all descended from therapsids, an order that seems to have been named by a palaeontologist with a lisp. It is a sign of the difficulty of assigning extinct species to the standard classification of life that the order is officially part of the reptile class, although in the living world mammals, which are descended from the Therapsida, are a separate class ranked alongside reptiles. But let that pass. Half-and-half creatures have to go in one category or the other, and since true mammals had not evolved by then, that leaves only one class for the therapsids to go in.

How, though, do you distinguish between a mammal and a reptile (or between a protomammal and a reptile), especially when you are dealing, not with living individuals in front of you, but with scraps of fossilized bone from rock strata hundreds of millions of years old? Most of us can list a few differences between mammals and reptiles off the top of our head. But be careful. Probably one of the first things that springs to mind is that reptiles (such as snakes) are cold-blooded. But Bakker has taught us that many reptiles may have been hot-blooded in the past (and since birds, which are hot-blooded, are descended from dinosaurs, which were descended from reptiles, there is some living evidence that he is right). Better leave that one off the list. So, what characteristics should we concentrate on? Mammals are covered in hair, while reptiles are scaly, but hair and skin leave little trace in the fossil record. Reptiles lay eggs, while mammals bring forth their young alive, and suckle them; but, although fossil eggs are known, the other relevant bits of biological apparatus do not fossilize. To the palaeontologist, bones are usually the only clues to work with, and there are several skeletal differences that make it possible to classify individuals as more or less mammalian. What they often come down to, surprising as it may seem, is the structure of the small bones in the ear.

Reptiles have a middle ear which contains one bone, used to

transmit sound vibrations inward. Mammals, including ourselves, have ears that contain three bones (hammer, anvil and stirrup), which compromise a more sensitive listening mechanism that works over a broader range of sound frequencies. In all probability, reptiles are literally incapable of appreciating Bach. Fossil remains show that this structure developed over a long period of evolutionary time, and that the two extra bones in the mammalian ear (hammer and anvil) are adapted from bones that are part of the jaw structure in reptiles. We hear, and appreciate, Bach with the aid of bones that reptiles use when chewing. Among other things, the nature of the ear bones revealed from studies of fossil skulls enables palaeontologists to tell whether an individual from the Permian or Triassic was more like a mammal or more like a reptile. They also, incidentally, provide a beautiful example of Darwinian evolution at work, and give the lie to ill-informed 'Creationists' who claim that there is no evidence for the step-by-step evolution of complex parts of the body, such as ears.

In fact, it is not so much that mammals evolved 'from' reptiles, but rather that both evolved from amphibians. In the late Carboniferous there was a variety of reptile forms around, but only two of them were to play a major part in the subsequent story of life on Earth. One group, the diapsids, kept what we regard as typical reptilian characteristics, and their line led to modern lizards, crocodiles and birds, as well as to the now-extinct dinosaurs and flying reptiles. The other group, the synapsids, did much better than the diapsids at first, but their descendants suffered an almost terminal setback in the Triassic, before seizing their opportunity, as mammals, when the dinosaurs disappeared.

Synapsids came in two main varieties – two orders. First, there were the pelycosaurs, which appeared late in the Carboniferous, shortly after 300 Myr, and had died out by the end of the Permian, some 50 million years later. They were members of Bakker's Megadynasty I, cold-blooded, lizard-like creatures with a sluggish lifestyle. These gave way to the therapsids in an explosion of evolutionary development following the 'invention' of internal warming mechanisms in the middle of the Permian. Their successors were the therapsids, which came in so many varieties that we cannot even list them here. The range included: creatures rather like heavy-set dogs, that probably browsed among vegetation in and around shallow water; large, cow-like forms whose remains are found in early Triassic rocks from South America, South Africa and Russia; bear-like carnivores a couple of metres long; and a species known as *Moschops* which had a very thick skull, and probably went in for head-bashing contests, rather in the way goats do today.

By the Early Triassic, a little after 250 Myr, the carnivorous *Cynognathus* was around. Individuals of this species were about as big as a modern badger, with a big head and a short tail; they were almost certainly covered in fur rather than a scaly skin, as was *Thrinaxodon*, a smaller, stoat-sized creature from the same geological period. The therapsid that most closely resembled a mammal came slightly later, and its remains are found in rocks from the Early Jurassic, just after 200 Myr. This was *Probainognathus*, a member of the sub-order of therapsids known as the cynodonts. It had a skull, jaw and teeth very like those of mammals, and, like *Cynognathus* and *Thrinaxodon*, members of *Probainognathus* were probably furry creatures; they may even have suckled their young. But they were much smaller than their two predecessors, and played only a minor role in the story of life during the Jurassic. Their descendants evolved into three kinds of early mammal, one of which went extinct, one of which was the forerunner of modern egg-laying mammals (such as the duck-billed platypus), and one of which, *Kuehneotherid*, was the ancestor of every other mammal alive on Earth today. Such an important future role for *Kuehneotherid*'s descendants would have been scarcely credible, however, for any alien zoologist visiting Earth at any time during the 175 million years from the beginning of the Triassic to the end of the Cretaceous. Large therapsids had been swept away, first by some natural catastrophe and then by the competition of the superior life forms that evolved from the other reptilian line, the diapsids, and became the dinosaurs. Bakker's Megadynasty III dominated the scene. Two hundred million years ago, you would not have put good money on the *Probainognathus* line one day inheriting the Earth.

The day of the dinosaur

The wave of extinctions that struck at the end of the Permian was the greatest in the fossil record. It was not just the end of a period of geological time, but literally the end of an era, the Palaeozoic. The Mesozoic era, which includes three periods (Triassic, Jurassic and Cretaceous) began with a bang, at 248 Myr. Just as the amphibians were washed ashore in the aftermath of Late Devonian extinctions, so the day of the dinosaur was preceded, and probably triggered, by the extinctions at the end of the Permian.

In the seas, the terminal Permian extinctions wiped out between 75 and 90 per cent of all existing species. They took place over an inverval of about 10 million years, just at the time when Pangea was assembling almost all of the land surface of the Earth in one mass stretching from the South Pole to the North. As with the extinctions at the end of the Ordovician and the Devonian, a variety of geological evidence reveals that the Earth cooled in the Late Permian, and the primary reason for this cooling was undoubtedly the way the changing geography of the globe altered the circulation patterns of warm ocean currents and allowed ice to spread over at least one of the polar regions. Because the Permian catastrophe was so severe, it is tempting to speculate that cosmic collisions may also have played a part. But whatever its cause, the cooling had a dramatic impact on life on land, as well on life in the sea. This was the first great extinction of land-based life. After the end of the Permian, the world looked like a different planet, as the surviving life forms diversified and adapted to the changed conditions.

At first, the therapsids seemed to recover well from the catastrophe, just as their ancestors had recovered from lesser extinctions during the second half of the Permian. Large animals had been wiped out in the disaster, while smaller creatures survived. This is a typical pattern in extinctions of land-based fauna, and can be readily explained if the disaster killed off large amounts of vegetation. Large herbivores would starve as their food supply withered away (even if they could survive the cold), and large carnivores would starve if there were no large herbivores to feed on. Smaller creatures which needed less food (and which could find shelter from the harsh weather) would have a better chance of survival. In the space of a few million years, the small protomammalian survivors of the terminal Permian catastrophe had evolved new lines of large grazers and carnivores. But at the same time, the other variety of reptiles, the diapsids, were adapting and evolving to the changing conditions.

While conditions were relatively stable, before the end of the Permian, the ecological niches for large animals were filled by protomammals, and no new lines could get a look in. When those large species were wiped out, both the small therapsids and the small, more reptilian diapsids had an opportunity to diversify. The result was a genuine clash between two different variations on the evolutionary theme, both starting out on an equal footing. As mammals ourselves, it is a sobering reminder that there is nothing intrinsically superior about the mammalian way of life (that we are not 'higher up' some hypothetical evolutionary ladder) to learn that our ancestors lost this equal battle.

Their reptilian rivals of the Early Triassic were the thecodonts, small animals about as big as a modern dog. Like the protomammals, they had limbs placed underneath their bodies, well able to support their weight and give them a good turn of speed. Some of them could run on two legs, balancing with the aid of a long tail held out behind them. They must have been warm-blooded for them to have competed so successfully with therapsids (quite apart from other evidence of a high metabolic rate, which Bakker has collated). As the line became increasingly successful, individual members of some species reached a weight of half a tonne. In an explosion of evolutionary success and diversification, thecondants were soon replaced by the archosaurs: forerunners of the dinosaurs, the crocodiles and the flying reptiles (the pterosaurs).

Bakker has described how, in the Early Triassic, the herbivore community was dominated by members of the therapsid order known as cynodonts. These were preyed upon by other varieties of cynodont, and, at first, just a few species of archosaur. Over the next 35 million years, up to the end of the Triassic, the pattern, as revealed by the numbers of fossils of different species found in rock strata gradually changed. First, the number of predator cynodonts declined, while the number of predator archosaurs increased. Up to the middle of the Triassic, the herbivorous cynodonts were still doing quite well, although after about 230 Myr they were being eaten not by their close relatives in the therapsid order but by their more distant archosaur cousins. At about this time, at least one branch of archosaurs became the dinosaurs, and the dinosaurs rapidly evolved both new varieties of carnivore and varieties of grazer. Not only were the remaining cynodonts being eaten by dinosaurs, but their food was being eaten by other dinosaurs, which must have been more efficient grazers than our closer relations, since by 200 Myr, in the Early Jurassic, cynodonts had all but disappeared from the scene. The survivors, the line from *Probainognathus* to *Kuehneotherid* to ourselves, got by by evolving into small, mouse-like creatures, too insignificant for dinosaurs to take much notice of, probably leading a nocturnal life style and living off insects and, perhaps, plants. The day of the dinosaur had arrived.

If you want to know more about dinosaurs, you will have to get hold of Bakker's book, and John Noble Wilford's *The Riddle of the Dinosaur*. From the end of the Triassic, 213 Myr, to the end of the Cretaceous, nearly 150 million years later, dinosaurs ruled the world. This is over twice as long as the time that has elapsed since the 'death of the dinosaurs'; our own line has been distinct from the lines leading to other African apes for no more than 5 million years, less than 4 per cent of the 'day' of the dinosaur (if the domination of the dinosaurs had literally lasted for 24 hours, then the equivalent lifetime of our line so far is about 48 minutes).

While dinosaurs roamed the Earth, Pangea broke apart, first into Laurasia and Gondwanaland, then into separate fragments which drifted around the globe and began to group together in new patterns. Flowering plants, although they had appeared in an early form in the Triassic, burst out into something like their modern forms halfway through the day of the dinosaur, in the mid-Cretaceous. Bakker argues that dinosaurs were directly responsible for the evolution of flowers, because their voracious munching of vegetation gave an evolutionary edge to plants that evolved better means of reproduction: to a plant, a dinosaur was part of the mechanism of natural selection (grasses, however, did not appear until after the dinosaurs had left the evolutionary stage). And at least four important extinctions struck life on Earth while dinosaurs dominated the scene. All of this, however, lies outside the scope of the present book. Just two factors are relevant to us now. What might have become of the dinosaurs, had the world not changed 65 million years ago? And what was it that wiped them from the face of the Earth?

Heirs to the dinosaurs

'Dinosaurs' never existed, at least not in the official scientific classification of animals. The term was invented by the British palaeontologist Richard Owen in 1841 to describe what he regarded as 'a distinct tribe' of reptiles identifiable from their fossil bones. The name comes from two Greek words, *deinos* and *sauros*, together meaning 'terrible lizard'; the popular image of dinosaurs certainly fits the name, although there were also placid, cow-like browsers that have to go by the name of dinosaur if all ancient reptiles are to have the name. Although Owen's suggestion caught the popular imagination, it never became part of the scientific classification. The term 'dinosaur' actually covers two distinct orders of reptiles, the Saurichia and the Ornithischia. These names derive from features of their anatomy, saurichians with more lizard-like hips, and ornithischians having hips more like those of birds. The crocodiles and the pterosaurs rank equally with these two orders of dinosaurs. The feature that saurichians and ornithischians share with each other (and with mammals) is that their legs were underneath their bodies, holding them upright and giving them more mobility than their sprawling reptilian cousins. Many dinosaurs, from both orders, could even stand and run on their hind legs.

Most – virtually all – palaeontologists accept the term 'dinosaur', even though it has no place in the strict scientific classification. Robert Bakker and his colleague Peter Galton, though, have gone a stage further. In the present classification system, the five classes are mammals, birds, amphibians, reptiles and fish. The two orders of dinosaur are divisions of the reptile class. The evidence that birds are descended from dinosaurs is, however, now very strong indeed, while the evidence that many dinosaurs were warm-blooded is, to say the least, persuasive. Dinosaurs, say Bakker and Galton, were more like birds than like any living reptile, so it is nonsense to classify them as reptiles. They propose revising the classification so that the five classes become mammals, dinosaurs, amphibians, reptiles and fish, with the class Dinosauria subdivided into not only the Ornithischia and Saurischia, but also the Aves (birds). Confusingly, modern birds are more like the 'lizard hipped' dinosaurs than the 'bird hipped' variety. The new classification has not caught on. We think this is a pity, not least because it highlights the fact that dinosaurs were much more like us, and other mammals, than like the reptiles we know today.

As Niles Eldredge, of the American Museum of Natural History, put it in his book *Life Pulse*, dinosaurs 'were the standard issue creatures of the day, and individual dinosaurs were as numerous as mammals are today'. The 'terrible lizard' image of the dinosaurs conjures up pictures of creatures like *Tyrannosaurus rex*, and even when we remind ourselves that there were herbivorous dinosaurs as well (*Tyrannosaurus* after all, had to eat something!), the first thing that springs to mind is an image of a kind of super-rhino, something like brontosaurus (*Apatosaurus*), 25 m long and built like a tank. But there were also dinosaurian equivalents of smaller mammals like wolves and goats. Chicken-sized dinosaurs must have eaten still smaller creatures (probably including our ancestors). Almost everywhere you look in the world today, if you see a mammal occupying an ecological niche and making a decent living out of it, you can be sure that there was a dinosaur equivalent filling much the same niche 100 Myr. There is one obvious exception: ourselves and the other primates. How come, if dinosaurs were so successful, that they never evolved intelligence?

The quick answer is that they had no need to. They got along very well without it, thank you, and the kinds of environmental pressure that gave rise to human intelligence simply did not exist in the Cretaceous. More recent environmental conditions, as we shall explain later, put intelligence at a premium and hastened the emergence of our own species. Things were different in the Cretaceous; but is that answer really satisfactory? It is hard to believe that intelligence would not be an advantage for some species in any environmental set-up, even though it might have emerged more slowly in more equable times. If so, it is even possible that intelligent dinosaurs might have appeared on Earth, sooner or later, had it not been for the catastrophe at the end of the Cretaceous that swept all large dinosaurs away, writing *finis* to the saga of Megadynasty III.

This 'might have been' has been developed by Dale Russell, of the National Museums of Canada, with the aid of a colleague, Ron Seguin. Their collaboration built upon Russell's 1970s study of the fossil remains of a dinosaur specimen which had a relatively large brain, walked on two legs, and had four-fingered hands with opposable digits (the fact that our thumbs can oppose each of the fingers on the same hand is what gives us the delicate grip needed for fine work, such as picking up grubs for food, chipping a flint to make an axe, threading a needle, or threading a nut onto a bolt). The ratio of brain to body weight of these creatures (with bodies somewhere between a turkey and an ostrich in size) placed them in the same range as large modern birds, or the less intelligent modern mammals. In 1977, astronomer Carl Sagan, in his entertaining and immensely popular book *The Dragons of Eden*, mentioned Russell's work and took things a stage further with the kind of speculative leap astronomers are famous for:

> If the dinosaurs had not all been mysteriously extinguished some sixty-five million years ago, would the *Saurornithoides* have continued to evolve into increasingly intelligent forms? Would they have learned to hunt large mammals collectively and thus perhaps have prevented the great proliferation of mammals that followed the end of the Mesozoic Age? If it had not been for the extinction of the dinosaurs, would the dominant life forms on Earth today be descendants of *Saurornithoides*, writing and reading books, speculating on what would have happened had the mammals prevailed?*

Russell has described the reaction to this suggestion in his contribution to the book *Dinosaurs Past and Present* (edited by Sylvia Czerkas and Everett Olson). There was, he says with masterly understatement, 'a certain level of curiosity about what highly encephalized dinosaurs might have looked like'. Russell

* Carl Sagan, *Dragons of Eden*, p. 135.

and Seguin satisfied that curiosity by building a model of the kind of creature that this dinosaur, which is also known as *Stenonychosaurus*, might have evolved into. They took its existing features, the large brain, upright posture and developing hand, and extrapolated their development forward into a mythical future in which the dinosaurs had not become extinct. The result, which they dubbed a dinosauroid, caused a sensation. Russell calculated that at the rate of evolutionary change that prevailed in the Late Cretaceous, a creature with a body weight the same as that of a modern human, and a brain to match, could have emerged within 25 million years (that is, at 40 Myr), and the dinosauroid was built to match that forecast. The dinosauroid model looks more like a human being than a dinosaur, and Russell argues that this form 'may have a nonnegligible probability of appearing as a consequence of natural selection within the biospheres of earthlike planets' because of the broad advantages conferred by, for example, the upright, bipedal posture (which frees the front limbs to develop hands capable of sensitive manipulation) and by the use of two eyes at the front of the head for steroscopic (three-dimensional) vision to look for prey and to focus on whatever those hands might hold. 'The humanoid form may be a special (nonrandom) solution to the biophysical problems posed by intelligence.'

Reaction to the idea has been mixed, as far as the experts are concerned, but some at least are favourable. Bakker, for example, says that 'one could quibble about details, but Russell is probably correct in general. Moreover, those large-brained dinosaurs were certainly clever for their time, and probably hunted the rat-sized mammals of the period' (*The Dinosaur Heresies*' p.372). In 40 million years of further evolution, *after* they had achieved the equivalent of human intelligence, would the descendants of those dinosaurs have gone on to develop space travel and voyage to other planets, or would they have invented nuclear war and destroyed the Earth? Such speculations belong to science fiction (and, indeed, have been entertainingly developed by Harry Harrison in his *West of Eden* trilogy). Whether, like Bakker, you agree that such extrapolation makes sense, or whether, like one palaeontologist we discussed this work with, you dismiss it as 'Dale Russell's fantasy', there is no doubt that it was a good thing for us that the day of the dinosaurs came to an end when it did.

If there had been no great changes to the environment 65 Myr, we would not be here now. Megadynasty II, dominated by protomammals and with more reptilian reptiles relegated to small-animal niches in the ecology, gave way to Megadynasty III, with the reptiles bursting out of their confinement and taking over, while mammals became small, scurrying creatures. Then, the roles were once again reversed in Megadynasty IV. If you feel bad about the way the dinosaurs treated our ancestors, take comfort in the revenge we take on the descendants of the dinosaurs evey time we eat an egg for breakfast. When all the larger dinosaurs were swept from the stage (and by 'large' we mean everything from about the size of *Stenonychosaurus* upwards; if the intelligent dinosaur had been just a little bit smaller, the story might have turned out differently), mammals were able to diversify and compete for ecological space in the niches left vacant. Probably because of the abilities they had evolved, under the pressure of natural selection, in order to survive at all in a world dominated by dinosaurs, this time around they were able to beat any opposition roughly their own size out of sight, and take over the animal world, with many larger species evolving in the process. But what did kill off the larger animals at the end of the Cretaceous?

A unique catastrophe?

During the age of the dinosaurs, old continents broke apart and the fragments drifted around the globe, colliding with one another and rearranging the geography until, by the Late Cretaceous, less than 100 Myr, the view of our planet from space would have been recognizably similar to the view seen by an astronaut today. The Atlantic Ocean was much less wide than it is now, and the Pacific was wider, with more freedom for warm water to circulate up into the North Polar seas. But it was recognizably 'our' planet.

The effects of all this on the forms of life that we find today were important. For example, at 130 Myr, although Gondwanaland was breaking apart there was still a land surface stretching from Africa and the Americas across Antarctica to

Australia (and well north of the South Pole, which was covered by water at the time). Marsupial mammals seem to have evolved in what is now Africa, and spread over this land bridge. In Africa itself they were replaced by the modern mammals, the placentals; but by the time this happened the bridge to Antarctica–Australia had gone. The two southern continents remained a safe home for marsupials, with no competition from placentals until man arrived on the scene and brought his animals with him. Which is why kangaroos and the like are found in Australia today, although their cousins in Africa were extinct tens of millions of years ago.

What matters from our own point of view, though, is how the changing geography of the globe began to alter the environment in the second half of the Cretaceous. Studies of changes apparent in the fossils of different species of sea-dwelling microorganism, the plankton, show the classic picture of a steady global cooling, with the extinction of many tropical species over an interval of several million years. Antarctica–Australia, breaking away from Africa and the rest of fragmenting Gondwanaland further north, was drifting towards the South Pole once again, with all that that implied. Many species were in decline well before the end of the Cretaceous, but many more disappeared very abruptly in a sudden extinction right at the end of the period.

It is the same picture on land. In several parts of the world, rocks from 65 Myr show traces of a *sudden* change in plant life, a change which fits the idea of a rapid cooling of the globe, right at the end of the Cretaceous. But the fossils also show a more gradual change in land-based life for millions of years previously. The dinosaurs themselves dwindled over a span of at least 10 million years, from 30 genera to 13 found in the fossil beds of Montana and southern Alberta. And the 'terminal event' was not all that terminal. Some researchers suggest that as many as nine species of dinosaur actually lived on into the next period of geological time, the Palaeogene.*

* In the traditional geological timescale, the Cenozoic era, which began at 65 Myr (and followed the Mesozoic era), is divided into the Tertiary period, from 65 to 1.8 Myr, and the Quaternary period, from 1.8 Myr to the present. This is rather unbalanced, and some palaeontologists now prefer to divide the Cenozoic into two more equal periods, the Palaeogene, from 65 to 24 Myr, and the Neogene, from 24

This decline may have been caused as much by a drying out of the continents as by the cooling that took place. In the middle of the Cretaceous, although the landmasses of the globe were getting closer to their present positions, the interior of modern North America and large parts of South America were inundated, and there was a huge shallow sea in the gap between North Africa and Eurasia, which has now dwindled to become the Mediterranean. All this water would have helped to give the land nearby an equable climate, like the climate of Britain today. But when sea level fell, as a result of tectonic activity, the shallow inland seas dried out and continental interiors, without the moderating influence of large bodies of water, became exposed to harsh winters and baking summers. This climate of extremes can have done no good to the vegetation or to the dinosaurs that fed on those plants. The picture painted by palaeontologists such as Steven Stanley is of a world in decline, with plants and animals, both in the sea and on land, suffering the conseqences of a major (but slow) environmental rearrangement, when something extra – meteorite impact, outburst of volcanism, or both – struck at the already weakened populations. Niles Eldredge, who has a knack for a snappy phrase, sums it up best: 'the role of an impact was to make a bad situation truly awful'. Without that last straw, perhaps the events of the Late Cretaceous would have amounted to no more than a lesser extinction, with dinosaurs recovering afterwards, as they had four times before.

There is indeed evidence that an impact *did* occur at the end of the Cretaceous, and this may have been the crucial event that tilted the balance against the dinosaurs and in favour of the resurgent mammals. The extinctions do not follow exactly the same pattern as the earlier ones we have discussed, and in particular there is some geological evidence that the world first warmed, then cooled at the dawn of the Palaeogene, while sea levels dropped dramtically.

There are enough detailed 'explanations' of these events

Myr to the present. We go along with this. Note that the terms 'Cretaceous–Tertiary', 'Cretaceous-Palaeogene' and 'Mesozoic–Cenozoic' all refer to the same marker in the geological calendar, at 65 Myr, when (most of) the dinosaurs died out. Since the first epoch of the Palaeogene/Tertiary is called the Palaeocene, in both classifications, the boundary can also be called the 'Cretaceous–Palaeocene'.

around to fill several books; we pick just one, as an example as much of human ingenuity in thinking up disaster scenarios as of the susceptibility of Cretaceous creatures to climatic change. In 1988, Michael Rampino and Tyler Volk, of New York University, linked several fashionable ideas together. They used evidence that 80 per cent of marine plankton was destroyed at the end of the Cretaceous. This is revealed by changes in the proportion of the isotope carbon-13 in carbonate rocks from that time. Levels of carbon-13 in those sediments show a drastic reduction in the amount of calcium carbonate deposited, over a span of at least 350,000 years. As calcium carbonate is chiefly the remains of dead plankton, this confirms that little plankton was around at that time.

Whatever caused the extinction of plankton – and Rampino and Volk favour the impact of a meteorite – there should have been severe consequences for the climate. Supporters of the Gaia hypothesis, the idea that environmental conditions on Earth are kept more or less in balance by the action of living organisms, have recently become interested in how plankton affect cloud cover. It sounds crazy – tiny organisms in the sea controlling *cloud cover*? – but it seems to make sense on closer inspection. Plankton produce large quantities of a substance known as dimethyl sulphide, or DMS. DMS gets into the air, where it reacts to produce the sulphur-bearing 'seeds' on which the water droplets that make up clouds can condense. Clouds, by and large, help to keep the Earth cool, by reflecting away incoming solar energy. Rampino and Volk calculate that removing 80 per cent of the plankton from the oceans would reduce cloud cover sufficiently to make the world warm by 6°C, while a plankton decline of 90 per cent would cause a warming of nearly 10°C. If that happened suddenly, as the result of a meteorite plunging into the sea, it would certainly disturb the balance of life on Earth!

It is an entertaining scenario, but not one we would want to push too strongly. (Less entertaining, but equally plausible, is the possibility that so much vegetation died and rotted as the inland seas dried out that the carbon dioxide concentration of the atmosphere rose and temporarily enhanced the greenhouse effect. The lack of carbonate sediments also tells us that carbon dioxide was not being taken out of the atmosphere; with no plankton around, the world would have warmed even without the DMS effect.) Nobody knows exactly what happened. The fact that something unusual, and perhaps unique, did happen at the end of the Cretaceous is, however, borne out by another study published in 1988.

The question taken up by Thomas Crowley, of the Applied Research Corporation in Texas, and Gerald North, of Texas A & M University, is whether climatic catastrophes in general can arise as a result of gradual changes, such as the drift of a continent over a pole, and whether the terminal Cretaceous event, in particular, fits that picture. They showed that you do not need an external cause – such as a meteorite impact – to produce abrupt environmental changes. There can be a discontinuous response, a sudden jump from one climatic state to another, as a result of slow variations in the background environment.

An example of this kind of change would be a steady increase, or decrease, in the carbon-dioxide content of the atmosphere. The kinds of sudden change that can result are familiar from a branch of mathematics known as catastrophe theory. They happen if a system – in this case, the climate of the Earth – can exist in either of two (or more) stable states for the same external conditions. For example, with a particular amount of carbon dioxide in the atmosphere, and therefore a particular strength of the greenhouse effect, the Earth might be stable either without any ice caps or with a polar ice cap. Computer simulations show that this is indeed possible for some global geographies (perhaps including, interestingly, the present-day geography of the globe): with no ice cap, the climate is stable; but if an ice cap forms, it can maintain itself by reflecting away solar heat. In such a situation the Earth might be ice-free, but with the potential for glaciation. If the carbon-dioxide concentration then slowly declines, a point will be reached where the ice-free state is no longer possible. At that point, the system must flip into the stable state with an ice cap. But, and this is the nub of the matter, even if the carbon-dioxide concentration now increases slowly once again to where it was before, the world stays glaciated because there is no trigger to switch it back into the stable ice-free state.

When the two Texas researchers put some numbers into their computer simulations, they found that a change of this kind can occur for a temperature change equivalent to a variation of just 0.0002 per cent in the Sun's energy output. This is well within the range of variations that could be caused by changes in the carbon-dioxide content of the atmosphere, or by changes in the ocean currents caused by the changing geography of the globe. Two of the great extinctions of the past 600 million years, at the end of the Ordovician and in the Late Devonian, fit this pattern extremely well, with a sudden onset of ice-age conditions following a long spell of ice-free climate; other extinctions we have mentioned fit the pattern reasonably closely. The extinction at the end of the Cretaceous, however, is quite different, because there is no evidence that the Earth flipped into a long-lasting glaciated state at that time at 65 Myr. The one big extinction where there is least evidence for this instability effect at work is the very extinction where there is reasonably strong evidence for an extraterrestial impact. It really does look as if the terminal Cretaceous event was special, and we can put this in an intriguing context.

At the time the dinosaurs disappeared, the environment was in decline and Antarctica was settling towards the pole. But the world had *not* yet entered an ice age. Suppose there had been no meteorite impact, or whatever, that made things awful instead of merely bad. Then, the dinosaurs might have recovered from the merely bad conditions. We know, from geological evidence, that a series of ice ages actually began about 5 Myr, as the present geography of the globe became established. It would have happened anyway, and probably brought an end to the age of the dinosaurs (unless they had developed intelligence and a technological civilization) without the meteorite impact. It seems, on this evidence, that the Cretaceous actually ended 60 million years too soon. By rights, and from the evidence of previous extinctions, it should have taken that long for the climate to deteriorate to the point where massive extinctions occurred. The unique events that occurred at 65 Myr – which, tantalizingly, we may never be able to understand fully – brought a premature end to the age of the dinosaurs, and gave mammals an early start back up the road to prominence. By the time the ice arrived, 60 million years after the dinosaurs' departure, the mammals were ready for it. And that, literally, is where we came in.

CHAPTER FOUR

The Return of the Magnificent Mammals

By 65 Myr, the main landmasses of present-day North America, Eurasia and Africa were beginning to become distinct continents, as the northern super-continent known as Laurasia began to break up. But the proto-Atlantic Ocean was still a relatively narrow sea, closed at its northern end where Greenland still snuggled between Europe and North America; so animals could move reasonably freely across at least the northern region of the supercontinent. In the south, the equivalent supercontinent, Gondwanaland, still covered the pole, with Antartica more or less in its present position, but with South America and Australia still attached to it. The continuing break up of Pangea, the old single landmass which had formed around 225 Myr, and the rearrangement of the fragments, was to have a profound influence on the changing climate of the globe over the next 65 million years, and on the emergence of humankind. During this 65-million-year period, the Cenozoic era, half of all the sea floor of our planet has been recycled, crunched out of existence beneath continents and born anew at spreading ocean ridges. Since the seas cover two-thirds of our planet, this means that one-third of the entire surface of the Earth has been renewed since the death of the dinosaurs.

The seeds of our species, however, were already in existence all that time ago, as the day of the dinosaurs drew to a close. We belong to the order of mammals known as primates. Primates evolved through adaptation to a life in the trees; they include about 60 varieties (genera) of living animals (among them lemurs, monkeys, ourselves and other apes), and perhaps twice as many extinct varieties known only from their fossil remains (plus an unknown number of extinct varieties that have left no fossil remains for us to study).

Life in the trees requires a different set of abilities to life on the ground, and over millions of years evolution has selected those advantageous characteristics in ourselves and our close relations. You need good physical agility and coordination in order to get about among the branches, and good eyesight in order to find food, look out for enemies and to see if the next branch you are going to grab is safe. Stereoscopic (three-dimensional) vision, so that you can judge distances and jump safely from branch to branch, requires your eyes to be at the front of your head, with overlapping fields of view. The ability to hold on to branches with all four limbs (and preferably a tail as well) is essential, and at least a couple of those limbs have to end in hands that are capable of holding food, either to carry it to a place of safety or while it is being eaten. Drop a tasty morsel from a treetop, and you have lost your lunch for good. The variety of food available – some fruit, some insects, the occasional piece of meat – affected the evolution of primate teeth. While other varieties of mammal specialized in developing either sharp cutting teeth (for meat eaters) or big grinding teeth (for vegetarians), our omnivorous ancestors kept the full variety of teeth found in the ancestral mammal forms (incisors, canines, premolars and molars) in a fairly neat all-purpose package. Finding the variety of food, making use of the good eyesight and coordinating the agile limbs in speeding through the trees all took a reasonably sophisticated nervous system, and so primates also evolved relatively large brains for their body size.

So, we have an identikit picture of a primate: a mammal with a large brain, good stereoscopic vision (and good hearing), with good grasping hands and all-purpose teeth. Another characteristic feature, which probably evolved from the need to hold small babies safely in the treetops, is that female primates have a pair of milk glands on the chest. The whole package is, hardly surprisingly, recognizably human. The surprise is that creatures whose fossil remains show enough of these features for them to be classified as primates already existed around 65 Myr, in the region that is now Montana. (But then again, if ancestral primates were around at the end of the Cretaceous, it is no surprise that most of their descendants, after a further 65 million years of evolution, are so superbly adapted to life in the trees today; the interesting thing is the way some of those descendants – ourselves – have abandoned the archetypal primate lifestyle.)

In those days, the climate of Montana was very different from what it is today. To start with, North America was closer to the equator. In addition, as well as there being a gap between North and South America, there was an open seaway between Africa and Eurasia. So ocean water could circulate freely around the globe at low latitudes, while on the other side of the world from Laurasia there was a huge ocean with unrestricted access to the North Pole. All of this helped to keep the climate warm and equable. But as the fragments of Laurasia drifted further north, and as this and other factors helped to cool the climate, primates became restricted to the tropical zone – not so much because they could not stand the cold themselves (after all, they were furry and warm-blooded), but because they needed an all-year-round supply of fruit and insects to eat.

The megayears of boom

Although the oldest known primate fossil comes from North America, remains almost as old have been found in Africa and South America, and nobody can say exactly where or when our order began. It began, though, in interesting times, maybe (as we have seen) in the aftermath of a cosmic impact, and certainly at a time when the Earth's magnetic field – possibly as a result of that impact – went crazy. Between about 80 and 70 Myr there was just one reversal of the field; in the 10 million years around the end of the Cretaceous there were at least 16 reversals. Nobody can be sure why this happened (though it does seem likely that it was associated with the break-up of the supercontinents), or what effect it may have had on the climate of the Earth or on life; interestingly (and perhaps even alarmingly, if you take a very long-term view), the frequency of reversals has increased, erratically, over the ensuing 65 million years, with roughly 40 reversals in the past 10 million years. This increase in magnetic activity has accompanied a long, slow cooling of the globe. That may be a coincidence, but the cooling, at least, has played a major part in our story.

Mammals in general, not just primates, participated in the recovery of life from the catastrophes, whatever they were, that had brought an end to the reign of the dinosaurs. Not just the Cretaceous period, but the Mesozoic era, had ended; in the Palaeogene period (the first of the Cenozoic era) there were new opportunities for the survivors to multiply, evolve and adapt.

At first, just about all they did was multiply. Mammals spread out, and increased in numbers, but stayed fairly small and rat-like. But within a few million years the first mammal boom was well under way. New species evolved to fill the niches left vacant by the demise of the dinosaurs, and many mammals grew in size and developed a bigger brain in order to do so. At the same time the supercontinents were breaking apart, so that some varieties evolved only in some parts of the world, and could not spread to other continents (the classic example is the evolution of marsupials across the southern supercontinent). Some of the early beneficiaries of the death of the dinosaurs were mammals that would look distinctly weird to modern eyes. There were grazers, for example, that grew to look like sheep the size of rhinos – but grazed on vegetation other than grass, which, although it emerged early in the Palaeogene, did not spread to dominate large areas of the world until the end of that period, about 25 Myr. Equally bizarre carnivores fed on these giant 'sheep'. They included the creodonts, predators with prominent skull-crests, some of which evolved into creatures bearing a superficial resemblance to the sabre-toothed tiger of more recent times. Mammals were again moving back into the seas from which their ancestors had come so long ago, producing several kinds of whale by the early Eocene, an epoch within the Palaeogene that began about 58 Myr. At around the same time, the continuing evolutionary activity on land saw the emergence of bats, true rodents and hoofed animals, including ancestral horses the size of modern dogs. By 50 Myr, ancestral elephants the size of pigs were also on the scene.

From our point of view, though, the most important evolutionary development during the Eocene was the emergence, about 50 Myr, of the first monkeys, a new variety of primate. Unfortunately, the fossil record of this event is sparse, and there is very little precise, direct information available about when and where the monkey line appeared. One of the most interesting features of this development, however, is that two kinds of monkey seem to have emerged separately, from the same kind of ancestral primate, in response to evolutionary pressures acting in the same way in different parts of the world.

By this time, South America was isolated from the rest of the continents except Antartica–Australia, where there is no sign that there were ever any monkeys. And yet, monkeys that look very similar to one another, and which occupy the same ecological niches in the forest, emerged both in South America and in Africa. The two groups – known today as 'New World' and 'Old World' monkeys – must have evolved independently from some earlier kind of primate, and yet their differences are quite superficial. The New World monkeys have developed a grasping – prehensile – tail, which they can use as a fifth 'hand', while the Old World monkeys can use their tails only as balancing poles; and there are differences in the nostrils of the two kinds of monkey. But those are the only two obvious ways to tell them apart. Their separate evolution shows that when the ancestral primate was set the evolutionary 'problem' of adapting to a particular environment, occupying an ecological niche in the trees of the tropical jungle and living off fruit and leaves, it came up with the same 'answer' twice. This is an example of what is known as parallel evolution. Within the 'gene pool' of primates – the DNA that we inherit from our ancestors – there is scope for a certain amount of variation and adaptation of the basic mammal form, and the kinds of change that will occur by chance and be selected by evolution operate within those limits to tailor the outward physical form of a species to its niche. For example, natural selection might favour primates with either more or less hair; but a primate line cannot suddenly evolve feathers in place of hair.

The idea of parallel evolution, that different species are tailored in similar ways by natural selection to fit their ecological niches, is very much the sort of argument that leads Dale Russell to imagine that an intelligent dinosauroid would look rather like a human being, or perhaps we should say, 'rather like a primate'. It also explains why some species of small whale and dolphin, mammals that returned to the sea relatively recently, resemble sharks, fish whose ancestors never left the sea. In order to be an efficient swimmer you have to be a certain shape, whatever your ancestors were. And if that still seems a little far-fetched, we can digress briefly to look at an even more striking example of parallel evolution at work, operating not just within the confines of the primate gene pool, but in the gene pool of the whole mammalian class.

Mammals that give birth to their young as, essentially, tiny embryos and keep them in a pouch until they are big enough to emerge into the world at large are called marsupials. Our kind of mammal, which gives birth to fully formed young that can, in many cases, run with the herd almost as soon as they are born, are called placentals. (Some primates, humans especially, are a special case, for their young are still relatively helpless at birth and need more time to develop, because of their large brains.) Placentals are a more efficient form of mammal, in terms of the evolutionary survival of the fittest. We do not have to worry about why this is so, but it is a fact that everywhere that placental mammals and marsupials have come into direct competition with one another, the marsupial form has died out. Particularly clear is the evidence for this from South America, where marsupial evolution continued successfully until about 3 Myr, when the land connection to North America became established. As placental species invaded from the north, marsupials were pushed back and have been completely wiped out in just a few million years.

Australia, though, remained cut off from the rest of the world, until very recent times when people have introduced some species of placental mammal to the island continent. It is full of examples of parallel evolution. The native Australian marsupials include the equivalent of placental mice and rats, animals called dasyures rather like cats (including the Tasmanian devil), koalas that resemble bears, and kangaroos that, although they look very different, play the same role in the ecology as deer and antelope. The range of variation allowed by the mammalian gene pool seems to have allowed both placental and marsupial mammals to respond to equivalent evolutionary pressures in much the same way. A few exceptions, like the kangaroo, have produced different physical forms to fit the same ecological niche. A dasyure looks like a cat for the same reason a dolphin looks like a shark: because it has to be that shape to make a living in the environment it occupies. Starting from the same kind of basic mammalian stock, and applying the same kind of evolutionary pressures, you generally get the same 'answers'.

But in spite of the fascination of delving into the story of the evolution of other lines, the only way to keep our present story manageable is to focus ever more closely on the line that leads to ourselves as we come closer to the present day. Among other things, that means ignoring what was going on in South America and Australia, and concentrating on the development of the primate order in Africa. For it was there that some monkeys became apes, and some apes became human beings. Old World monkeys were in place in that part of Africa that is now the Sahara Desert, but was then lush forest, between 50 and 40 Myr. They lived in the northern part of what was then an isolated continent, with open sea between Africa and Eurasia. As both continents moved north, Africa was catching up with Eurasia, and would eventually collide with it, welding the two continents together in a union that has persisted to the present day. But before that happened, the ape line had already evolved from monkey stock; and the whole mammal class had been through an upheaval of extinctions caused by a series of catastrophic events and a major change in the pattern of ocean currents and the weather of our planet.

Setback – and recovery

The crisis began about 40 Myr and lasted for about 8 million years, covering the later part of the Eocene and the early Oligocene, which began about 37 Myr. A lot of evidence for changes around this time comes from studies of the fossil remains of creatures that lived (and died) on the sea floor, such as molluscs. Extinctions were mainly of species and genera, not whole families and orders, but were still substantial. The most detailed information about the changes that took place around this time comes from studies of microfossils, the remains of the tiny plankton that live in the oceans and whose chalky shells build up in the sediments on the sea floor. They show that, apart from being less extreme, the changes in plankton around 40 Myr followed a very similar pattern to the extinctions at the end of the Cretaceous, species that liked warmth disappearing from the scene. Just 2 million years later, a second wave of extinctions swept through the surviving plankton species, and there is also geological evidence for an asteroid or comet striking the Earth at about this time. The end of the Eocene epoch may well have been brought about by a similar, though smaller, cosmic catastrophe to the disaster that struck the dinosaurs at the end of the Mesozoic era. But this cannot fully account for the changes that took place in the early Oligocene, because there were still three more waves of extinction to come, culminating about 32 Myr. By and large, species declined over a long period of time, the recognized extinction 'events' being merely the culmination of a deterioration that was happening anyway. The whole pattern of changing life in the ocean, over the best part of 10 million years, can best be explained by global cooling on a dramatic and sustained scale. In round terms, studies of changes in different kinds of marine organism that lived at different depths in the sea show that the average temperature of surface waters in the tropical Pacific Ocean fell from 23° to 17°C, while the temperature of deep ocean water fell from 11°C to 5°C. And the effects of the global cooling show up in the traces left by life on land as well.

Temperature changes on land are recorded by changes in fossil plant remains: seeds' spores, fruit and pollen, as well as leaves and larger pieces of plant. Before the changes at the end of the Eocene and in the early Oligocene, regions at all latitudes of the world, it seems, enjoyed the kind of climate we regard as tropical. But during this 10 million years of cooling, many tropical and subtropical species disappeared from higher latitudes. Forty million years ago London was further south than it is now, at about the latitude of Madrid, or Washington, DC, today. That helped to keep it warm, but still placed it some 40° north of the equator. The deposits known as London Clay, in southern England, contain a detailed record of the changes that occurred between about 40 and 30 Myr, and show that, although at the end of the Eocene the vegetation there resembled that of the Malaysian jungle today (on the equator at 0° latitude), by 30 Myr the pattern had changed to that of a temperate climate, still warmer than today, but with seasonal variations. Think of tropical jungle in Philadelphia today, and you have some idea of the magnitude of the climate shift that occurred at the end of the Eocene and into the Oligocene. Seasons, in fact, could be said to have been invented at about this time. For hundreds of millions of years previously, up to the end of the Eocene, there had been little difference in climate at different latitudes, and even at high latitudes there had been little variation in the weather pattern during the course of a year. But by 30 Myr winters at high northern latitudes were distinctly colder than summers, with frosts, if not snow, becoming common. Even more dramatic changes were taking place, as we shall see, in Antarctica.

Life on land seems to have suffered two main extinction crises around this time, one at the end of the Eocene and one at about 30 Myr. Mammals were particularly affected by the first extinction, with many of the weird and wonderful varieties that had emerged in the post-dinosaur wave of mammalian evolution being wiped out, eventually to be replaced by newer and (to us) more familiar forms. (To put some of this in perspective, it helps to recall that the events we are now describing happened before 30 Myr, about halfway between the present day and the death of the dinosaurs; the first wave of post-dinosaur mammals had a run on the evolutionary stage almost exactly as long as the run the second wave of mammals has now enjoyed.) Recognizable ancestors of giraffes, pigs, deer, cattle, camels and rhinos all made their debut in the Oligocene. Their emergence and evolution was influenced by the continuing changes in climate, and not just in temperature. As the world had cooled, sea levels had fallen, leaving continental interiors farther from the sea and with a less plentiful supply of rain. With less rainfall, in many parts of the world thick forest gave way, at last, to open grassland. The world began to look like it does now after about 30 Myr, and the animals in it, especially the mammals, began to look like modern animals. As we have already mentioned, one of the modern lines that emerged from these upheavals was the ape line, descendants of the Old World monkeys that lived in northern Africa.

The monkey line itself continued to enjoy its old lifestyle, high in the trees, even though the size of the area covered by dense forests in northern Africa was shrinking. If the climate had not changed, and the forests had not shrunk and been replaced in some regions by more open woodland, there might never have been any ape-like variation on the monkey theme. After all, look at what happened to the New World monkeys, in South America: nothing. Because South America stayed an isolated continent straddling the equator and drifted mainly westward and only a little north for tens of millions of years, lush jungle remained the order of the day, and the New World monkeys, superbly adapted to life in lush jungle, stayed as they were. Like the dog that failed to bark in the night, this is a persuasive piece of evidence, and lends weight to the view that it was the environmental changes in Africa that encouraged – or forced – some Old World monkeys to adopt a new kind of life style, one which can be characterized, even for these earliest apes, as being built around a more omnivorous diet than that of fruit- and leaf-eating monkeys. Even in Africa, where the lush jungle remained so did the monkeys; it is where the regular supply of tropical foodstuff began to disappear that some erstwhile monkeys, the evidence of fossil teeth shows, developed the ability to eat other foods.

There is other, circumstantial evidence that the climate-related upheavals of the Oligocene were responsible for the emergence of the ape line. The fossil skull which has the distinction of being recognized as the earliest-known remains of an ancestral ape comes from a region known as the Fayum Depression, in Egypt; in honour of this it is known as *Aegyptopithecus*. It was found in strata 28 million years old, dating from just after the upheaval, in the heyday of the second wave of mammals.

The 'first' member of the ape family (strictly speaking, still a pre-ape, not a genuine ape) was a cat-sized creature, a tree-climber with a supple back and long limbs, each ending in a grasping hand; it had a larger brain, compared with its body weight, than for any other mammal known from that time. Of course, all this information was not gleaned from a single skull; other remains of *Aegyptopithecus* and its near relations have been found in slightly younger deposits, and by about 25 Myr there were several types of proto-ape in northern Africa. Apes are distinguished from monkeys, in living species as well as in the fossil record, by several features. Apes tend to be larger than their monkey contemporaries, and although both climb in the trees monkeys run about on the top of branches, while apes swing along beneath branches (this shows up in differences in the structure of their skeletons, especially shoulders and hips, which are adapted to different forms of locomotion). Monkeys and apes have different types of teeth, reflecting their different diets; and apes have bigger brains and are more intelligent. There are other differences, but these will do to be going on with; they are certainly sufficient to pick out ape fossils from monkey remains, even in strata 25 million years old.

By now, we are talking about very close relatives of ours, members not just of the same order as ourselves (primates), but of the same family (Hominidae, or hominoids). The hominoid family, in fact, contains only humans and other apes (the name *Hominid* refers only to ourselves and any ancestors that we do *not* share with other living apes); it took about 20 million years, after the emergence of monkeys, for the first apes to split off from the monkey line, and it was not until some 25 million years after the arrival of *Aegyptopithecus* on the scene that the line leading to *Homo sapiens* split from that of our sibling apes. This is another pointer to the importance of environmental, and especially climatic, changes in our evolution.

Apes appeared in Africa from the monkey stock at about 30 Myr because their environment changed; but then nothing much happened to the environment for more than 20 million years, so there was no pressure for any striking new development in the ape line. Steady evolution and adaptation to the new world that had emerged from the Oligocene upheavals was the order of the day, in our line as in others. The world had by now changed so much that geologists set the end of the Oligocene epoch, and of the Palaeogene period, at 24 Myr, and from then on we are in the Neogene, the second period of the Cenozoic era. The first part of the Neogene is called the Miocene epoch, a relatively long sub-division of geological time which ended only 5 Myr. The early Miocene, in particular, was the time when mammals came into their own. There were continuing geological changes, including the upheavals that created the Alps, and activity under the eastern part of Africa that was lifting a large area like a dome and would eventually crack it open to produce rift valleys littered with lakes and edged by active volcanoes. The steady drift of continents to different latitudes continued all the while. All this brought about a continuing shift to drier weather in many regions, and an expansion of grassland at the expense of forest; the new prairies were inhabited by a greater variety of mammals than has been seen on Earth either before or since. Our class spread into every part of the globe, in great waves of migration, while our family diversified into several varieties of ape, some emerging from Africa to populate large areas of Eurasia as the two continents collided and were joined together. If any interval of geological time fits the description, this was when the world resembled the Garden of Eden. The new mammalian world, including the world of the apes, owed its existence to the final break-up of Gondwanaland, far away from Africa, and to the development of an ice cap over the South Pole as a result. Not just people, but all apes, are indeed children of the ice.

Winter in Eden

It was the cooling and drying of the globe in the Oligocene that created conditions that we regard as ideal, an Eden-like paradise on Earth. Such conditions seem so appealing to us because we have evolved and adapted to fit those conditions, just as the dinosaurs of old evolved and adapted to fit a more 'tropical' planet. Had they been around in the Miocene, they would probably have hated it. But it was the invention of winter that made Earth an Eden for our family.

There is no mystery about either the timing or the cause of this change in climate. The changes evident in fossil remains of plankton in cores drilled from the ocean floor and brought to the surface for analysis show that the mass of cold, deep water that is a feature of the circulation of the oceans today only began to develop at the end of the Eocene, about 37 Myr. The deep, cold layer developed because the Antarctic continent, over the South Pole, began to freeze about then. Very cold surface water at high southern latitudes would have been denser than the warmer water, and would have sunk down into the depths of the ocean before beginning its journey northward as a cold bottom current (just as hot air rises, so cold water sinks; and for the same reason: colder fluids are denser than their warmer counterparts). The cold water sinking near the pole and moving to lower latitudes cannot stay deep for ever. The pressure of more cold water piling in behind it eventually forces the current to well up in warmer parts of the world, bringing cold water to the surface and cooling even the tropics. Then, the cool surface water warms in the heat of the Sun as it begins its journey back towards the polar regions to complete the cycle. Such currents, originating near both poles, are a feature of the ocean circulation today; but before the end of the Eocene, when the whole world was warm, they simply did not exist.

Independent evidence of the timing of the growth of glaciers in the Antarctic matches with the timing inferred from studies of plankton from the tropical Pacific sea floor. There is some evidence that ice began to build up as early as 40 Myr, just when the series of Eocene/Oligocene upheavals began, and firm evidence that a full-scale ice sheet had developed in East Antarctica by the early Oligocene, at 35 Myr. The evidence comes from sediments of these ages that contain typical products of glacial activity – mixtures of clay, silt, sand and gravel – and which are dated by magnetic techniques. The outer edge of the ice sheet may have been as much as 140 km beyond the present ice front, and the evidence from eastern Antarctica is mirrored by similar traces of glacial activity from the Ross Sea, on the western side of the continent. Whether or not cosmic impacts added a 'last straw' effect for some or all of the extinctions that took place around this time, it is certain that Antarctic glaciation was the main cause of the change in climate and the long, slow decline of many species, including most of the first wave of post-dinosaur mammals. But why did Antarctica cool?

Before the end of the Eocene, Antarctica was still attached to both South America on one side, and Australia on the other. Warm ocean currents that flowed poleward from the tropics swept southwards along the edges of both Australia and South America, cooled only a little as they moved past the edge of Antarctica itself, and then turned northwards again to return to the tropics. But at the end of the Eocene, Australia and

Antarctica broke apart and Australia began to shift northwards, with its cargo of marsupials, to its present location (where it is not fixed, of course, but continues to drift to the north). This allowed a new, cold current to develop, flowing through the widening strait between Antarctica and Australia. This cold current deflected the southward flow of warm water before it could reach Antarctica, building up a barrier between the southern continent and the warmth of the tropics. Now, the main current flowed southwards down the eastern seaboard of South America, cooled at high latitudes, and moved eastwards almost around Antarctica, through the strait between Antarctica and Australia, before returning northwards up the western side of South America. With this barrier current in place, even though South America was still attached to Antarctica, the landmass over the polar region began to cool and ice began to build up, first over the land and then over the sea itself.

The growth of the glaciers may have been triggered simply by a slow cooling of Antartica until some critical threshold was reached. Or, perhaps, one of these suspected early Oligocene impact events may have cooled the southern hemisphere enough for snow to begin to lie in a permanent layer over the continent. The cooling may have been hastened by minor shifts in the position of Antarctica, which placed the continent more centrally over the South Pole. Whatever the exact cause of the spread of the first snowfields, though, once snow did begin to lie over the land a self-perpetuating southern ice age would have begun. A white, reflective layer of ice and snow bounces heat from the Sun back out into space, making the polar region much colder than it would be if there were no ice present, and encouraging the ice sheet to grow, and to chill the waters at the edge of the continent. Cold water could now sink into the depths to create the cold, deep layer of the ocean. This would have been a sudden change, whatever the exact trigger, and would have been disastrous for any inhabitants of Antarctica.

It is interesting to speculate on what kind of animal and plant life might have lived there before the ice began to spread. After all, even without permanent ice cover the south polar region still experienced a long winter night, during which no plants could grow and there would be no food for animals. Perhaps there was highly seasonal vegetation, eaten by animals that migrated southwards each spring and retreated towards Australia or South America, following the Sun northwards, each autumn. But any evidence for their existence, and for a massive wave of extinctions in that part of the world when the ice arrived, is buried now beneath glaciers many kilometres thick. Some survivors may have emigrated to South America; on the Australian side of Antarctica, however, there was nowhere left to go, unless you were a strong swimmer.

In the middle of the Oligocene, around 33 Myr at the time of the last great wave of Oligocene extinctions, geological activity deepened and widened the gap between Australia and Antarctica, making the cold current stronger and reinforcing the development of cold bottom water flowing northwards, ultimately to well up at lower latitudes. As well as evidence for the development of large ice sheets on land, the sea-bed cores from those times show that plankton adapted to cold conditions proliferated in the waters around Antarctica from then on. About 30 Myr, just at the end of the period of Oligocene upheavals, the last connection between Antarctica and South America was severed. Antarctica itself was not immune to continental drift, but its main motion at that time (and since) was a slight rotation, an anticlockwise swing around the pole. At the same time, South America was (and is) moving westwards. But the rotation of Antarctica was not fast enough to keep up with the drift of South America, and over millions of years the piece of land joining the two continents was stretched and bent, eventually breaking like a piece of stretched toffee. You can still see traces of the stretching and distortion of the rocks of the Earth's crust in any atlas which shows the region of the southern tip of the Andes and the horn of Antarctica; both westward-shifting pieces of continent have left trails of little islands behind them at the break, forming the Scotia Arc, but with the Antarctic horn lagging behind the tip of South America.

Once this final break with the rest of the world had occurred, Antarctica was surrounded by sea, and the current flowing eastwards along its borders became a true circumpolar current, circling the entire globe. Looking down from above the South Pole, this is a great clockwise current. To the north, from this perspective the surface layer of the ocean circulation system is dominated by three great counterclockwise currents, in the South Atlantic, the South Pacific and the Indian Oceans. The southern edges of these counterclockwise currents blend in with the northern edge of the circumpolar current, like the teeth of three great counterclockwise gearwheels meshing with the teeth of a clockwise-rotating cog. Once the pattern became established, the whole circulation of the southern oceans acted to reinforce the strength of the circumpolar current and to strengthen the barrier, in the upper layers of the ocean, between the cold water around Antarctica and the rest of the world. Surface water could get through to the edge of Antarctica only in the form of eddies, the oceanic equivalent of low-pressure weather systems; and once it got there and cooled; it could return to the tropics only in the form of a deep, cold current.

The cooling of the south polar region had a direct influence on the climate of the globe, enhancing the difference between winter and summer and cooling the world generally as cold winds, as well as cold ocean currents, moved northwards away from the pole. The root cause of this cooling was the fact that some solar heat that used to be absorbed by the Earth was now being reflected away by the ice sheets at the pole. But there must also have been an indirect, but perhaps even more important, contribution to the creation of winter in Eden.

The warmth of the world in the 20 million years or so following the recovery from the death of the dinosaurs really is astonishing when compared with present-day climate or with the climate of the Miocene. Malaysian jungle in England, even allowing for England being a few hundred kilometres closer to the equator, is not something you would expect to develop today even if the polar ice caps were removed. Nor, come to that, does it seem entirely plausible that a continent covering the South Pole, and subject to months of winter darkness, could remain entirely ice-free, whatever the ocean currents were doing. And yet, the geological evidence shows that it did, for 20 million years and more. Could some other effect have been making the world warm in the first part of the Palaeogene? Almost certainly, yes. The obvious candidate is the carbon-dioxide greenhouse effect, the process which traps in the lower atmosphere heat that would otherwise be radiated away into space by the warm surface of the Earth. When the atmosphere contains more carbon dioxide, then the world (other things being equal) is warmer. And, as it happens, the warming effect is stronger nearer the poles, where the input of heat from the Sun is weaker. The very richness of the vegetation in the early Palaeogene hints that there may have been more carbon dioxide around then, since as well as sunlight plants need carbon dioxide for photosynthesis, and thrive when there is more of it available. The warmth of the world in those days makes it almost certain that there was a thicker blanket of carbon dioxide then. Since carbon dioxide is produced by volcanic activity, there is no problem in explaining why there was more carbon dioxide in the air at a time when supercontinents were breaking apart, being torn at the seams along great rift valleys edged with belching volcanoes.

Today, although things are quieter on the geological front, natural volcanic sources still produce carbon dioxide (and so do human activities, as we discuss later), but the amount of the gas in the atmosphere has stayed roughly the same for millions of years because natural processes take carbon dioxide out of the air and lay it down in sedimentary rocks, in the form of carbonate compounds. Those natural processes are driven by a kind of oceanic pump (with biology playing a part) in which carbon dioxide dissolves in cold water at high latitudes and is carried down into the oceans, where plankton absorb it and use it to build their chalky, carbonate shells. When the plankton die, the shells fall to the sea floor, and gradually build up into layers of carbonate rock. Cold water is much more efficient at dissolving carbon dioxide than warm water, and this combined oceanic—biological pump is most effective at taking carbon dioxide out of the air and sequestering it in rock today at high latitudes, especially in the cold waters around Antarctica, which are rich in nutrients and where plankton thrive. But those cold waters were not there in the early Palaeogene. Before the cold Antarctic waters developed, the carbon-dioxide pump must have been less efficient; but when the flow of cold, deep water out of Antarctica became established, and the cold-loving plankton began to proliferate there, there must have been a sudden (by geological standards) drawing-down of carbon dioxide from the atmosphere, and a reduction in the strength of the greenhouse effect. As well as more solar heat being reflected away from the surface of the Earth, less heat was being trapped by the atmosphere on its way out into space. No wonder the world cooled in the Oligocene. The cooling was to continue, although much more slowly, right up to modern times. Although the mammals that proliferated in the Eden of the Miocene had no way of knowing it, it was to be downhill all the way, climatically, from then on.

The day of the ape

The mammal heyday of the Miocene was also the apes' heyday. *Aegyptopithecus*, the apes' precursor, emerged from the upheavals of the Oligocene shortly after 30 Myr, but the fossil evidence reveals very little about what was happening to our close relatives over the following 10 million years. After about 25 Myr, however, there seems to have been an almost explosive radiation of ape varieties, and by now these were true apes, with all the important distinguishing characteristics that we have mentioned. The success of the apes over the next 20 million years or so coincided initially with a time when the world warmed slightly, compared with the Oligocene, but remained dry. The dry conditions, with a lower sea level than before, were of course directly related to the establishment of a large ice cap over Antarctica. The size of the ice cap must have varied during the Miocene, but it seems that ever since the Oligocene there has always been some ice over the South Pole.

With the arrival of true apes on the evolutionary scene, this is probably a good place to stop and take stock of what evolution is all about. Many people, even today, have a vague picture of evolution as a climb up a long ladder, with ourselves at the top. The funny little cat-sized *Aegyptopithecus*, and the apes of the Miocene, might figure in that picture as imperfect creatures struggling to achieve the perfection represented by humanity. Stated that baldly, it sounds silly, and it is. Although the creatures whose fossil remains we study may have vanished from the evolutionary stage and gone extinct millions of years ago, they were well fitted to the environment in which they lived. Each creature in our own ancestral line was a success story in its own right, adapted to a certain lifestyle and occupying its own niche in the ecology. Indeed, there is no reason to single out the human line as special, except for our chauvinistic interest in it. All the rest of the variety of life on Earth in the Miocene, or before, was an evolutionary success story in its day, just as, indeed, termites, budgerigars, salmonella bacteria and all the rest are equally as successful as we are today, by the only evolutionary criterion that matters; the fact that they are alive and able to reproduce. There is no way in which we can claim to be 'better' than *Aegyptopithecus*, or the Miocene apes, only different. They were well adapted to the world in which they lived, and we are well suited to the world in which we live. The reason why we are here now, and not them, is primarily because the world changed at the end of the Miocene and created new evolutionary opportunities. The story of how and why that change happened will be told in the chapters that follow. But first, we want to bring the story up to the end of the Miocene, 5 Myr, and introduce the variety of apes that enjoyed the cool, dry conditions, with widespread open forest and extensive grasslands, of that epoch.

The closer we come in the evolutionary story to the emergence of our own species, *Homo sapiens*, the more the experts argue about the exact interpretation of each fossil and how it relates to ourselves. Fortunately these are very often only arguments about details, and about who should have the honour and fame accorded to the discoverer of a 'missing link' or the 'oldest human ancestor'. The arguments do not obscure the underlying picture of human evolution, and there is no need to worry about the details of the debate here; but if you are interested in them we recommend Roger Lewin's entertaining book *Bones of Contention*. Out of all this debate, a relatively new understanding of the emergence of the human line has developed over the past 10 years or so (a story detailed in *The Monkey Puzzle*, by John Gribbin and Jeremy Chergfas); from now on, we will be following this modern interpretation of our ancestry, and especially the summing up in another book by Roger Lewin, *Human Evolution*. Some of the names and dates we mention may not quite fit with the story of human evolution you have read about in older books, or which you learned about in school, or even from Richard Leakey's popular TV series of a few years ago (Leakey himself has changed his views on the timing of key steps in the emergence of humankind since that series was made). But the overall picture we present here is an outline of the best and most complete understanding of the making of humankind that science can now provide.

The day of the ape began at the beginning of the Miocene, at 24 Myr. By 20 Myr there were plenty of apes around in Africa, members of different species but grouped together by palaeontologists under the overall name of dryopithecines; 'pithecus' means 'ape' in Greek, and the prefix tells us that the creatures lived among the trees. The explosive radiation of the apes occurred a little later, when Africa nudged up against Eurasia and created a land bridge between the two continents. Apes and relatives of modern elephants, among other species, spread out of Africa and into Eurasia; cats and horses, among others, moved the other way. There may have been changes in climate that favoured the apes' way of life around this time, or it may simply have been that they found Eurasia to their liking. Whatever the reasons, although monkeys also moved into Eurasia following the collision between the two continents around 18 Myr, the apes diversified much more and spread into more ecological niches. In terms of species, around 15 Myr apes outnumbered monkeys by 20 to 1, judging from the fossil evidence, and there is neither any point nor any need to elaborate on all the places they went or all the varieties that developed. It is enough to say that apes were one of *the* success stories of the Miocene, and that the present handful of ape species on Earth, including ourselves, represents a pathetic rump of a once great family.

There are just three groups of apes from around this time that are of interest to palaeoanthropologists. One is known as *Gigantopithecus* (from the size of the fossil remains), and the other two as *Ramapithecus* and *Sivapithecus*, both named after Hindu gods. Together they are sometimes known as 'ramamorphs'; they were sufficiently like the modern apes for *Ramapithecus*, in particular, to be thought for a time to be our own ancestor, the first step along the line leading to human beings from the basic ape stock. Remains of *Ramapithecus* have been found in deposits as old as 14 million years and as recent as 7 million years, so it was certainly a successful variation on the ape theme. For comparison *Homo sapiens* proper has been around for less than 100,000 years, and the uniquely *Homo* line, we now know, for about 3 million years; in terms of longevity, as yet we are less than half as successful as *Ramapithecus*. But the idea that *Ramapithecus* was uniquely an ancestor of ours, and not of other modern apes, has now been rejected. A combination of recent fossil finds and new techniques for analysing the closeness of the relationship between living species (including humans and other apes) from studies of their DNA show that the split between ourselves and the other African apes was even more recent, not happening before 5 Myr. *Ramapithecus* is now thought to have been the ancestor of modern Asian apes, such as the orang-utan. These are our evolutionary cousins, but we are even more closely related to the African apes, the two kinds of chimpanzee and the gorilla. Indeed, by the usual classifications, people *are* African apes, although we like to think of ourselves as something special, in a category of our own, the hominids. ('Hominid' is a term we reserve for ourselves and for any ancestors of ours that lived after the evolutionary split with our nearest ape relations; 'hominoid' is a broader category that includes other living apes and their immediate ancestors.)

Unfortunately, none of the many varieties of ape around in the Miocene can be confidently identified as the ancestor of the African apes. Although some members of the ramapithecine line stayed in Africa while others migrated to Asia, through the Miocene forests that stretched across the newly joined northern landmass, the African branch of the family, according to the consensus view among palaeoanthropologists today, seems to have died out, while some other unknown form of ape was flourishing, or at least *surviving* to leave descendants, some of whom were our own ancestors. The vagaries of the fossil record mean that we know of a variety of creatures that were ancestral to both ourselves and the ramapithecines around 20 Myr (the dryopithecines), and we have fairly good evidence about relatives that may not have been our direct ancestors but who lived from 14 to 7 Myr (the ramapithecines); but in that standard picture none of the direct ancestors of the human line seem to have been preserved in fossil remains from about 18 Myr all the way down to 3.6 Myr. We have our own doubts about this standard interpretation, which we shall air in Chapter Five; but the picture becomes much clearer after

Ramapithecus has left the stage. The evidence from 3.6 Myr is for a distinctly hominid, not merely hominoid, ape, the earliest known direct ancestor of our own distinct line, separate even from that of the other African apes. However, all the signs are that the split from the lines leading to the chimps and gorilla had only just occurred at that time, and that it had occurred because of another upheaval in the Earth's climate.

Downhill all the way

When Africa and Eurasia were welded together around 18 Myr, it marked the beginning of the end of the pause in the climatic deterioration, a pause that had made the first part of the Miocene marginally more comfortable for life on Earth than the Oligocene had been. Ocean currents flowing between the two landmasses had helped to maintain equable conditions, but with this flow cut off the climate in the heartland of Eurasia and in northern and eastern Africa became drier and more seasonal. This may have helped the early dispersal of the apes out of Africa, providing them with the kind of open woodland (as opposed to tropical jungle) that their lifestyle was suited to; in the long term, however, as the trend continued and many regions became too dry for trees, the changes would bring an end to many of the varieties of ape that flourished in the Miocene, and indeed to many other species of mammal as well.

The pattern of life in the oceans changed significantly around 14 Myr. Although there had been ice at sea level somewhere around the southern continent ever since 35 Myr (or even earlier), the amount of ice cover had varied; at 14 Myr, though, the Antarctic ice sheet expanded, and stronger deep, cold currents pushed their way northwards. It has remained fully glaciated ever since. With more water locked up in the great ice sheets the world became drier still, and grasses came into their own in semi-arid regions of the globe. At around the same time volcanic activity increased, perhaps because of increased tectonic activity, and there is evidence of a moderately large cosmic impact in Europe. Many species of mammals, and even genera, disappeared at about this time, and most of the dryopithecine apes went extinct. But, as always at a time of evolutionary upheaval, some varieties survived and adapted. It may be no coincidence that the earliest-known ramapithecines, at 14 Myr, date from around this time.

There were few dramatic changes in climate over the next 10 million years, and no major extinctions to rank with some of the catastrophes we have discussed, but a succession of relatively minor events mark the decline in climate. Mountain glaciers appeared in Alaska around 9 Myr, and left their traces in the form of scars in the rocks; at a little less than 7 Myr, glaciation spread to the mountains of South America; and by 6.5 Myr there was so much water locked up in Antarctic ice that sea level fell below the level of the neck of land joining Africa and Europe across the equivalent of the present-day Strait of Gibraltar. Over an interval of more than a million years, the entire Mediterranean Basin repeatedly dried out and refilled. The Mediterranean Sea today loses about 3300 km^3 of water by evaporation each year, and is maintained only by the inflow of water from the Atlantic; without that inflow, it would take just a thousand years to dry out, leaving a layer of salt tens of metres thick; and, since the water evaporating there must fall as rain somewhere else, raising the average sea level around the world by 12 m. Cores drilled from the bed of the Mediterranean show several layers of salt produced in this way, evidence that the sea dried up several times. On each occasion, when sea level later rose sufficiently for water to break through again (presumably when the Antarctic ice retreated slightly), water would have thundered in at a rate of 40,000 kim^3 a year, 10,000 times faster than the present flow rate of the Niagara Falls, taking a hundred years to refill the basin. Geological activity, opening and closing the gap between Africa and Spain, may also have played a part in this cycle of Mediterranean droughts.

Whatever the exact cause of the desiccation of the Mediterranean, these changes must have had a profound effect on life in the region. When the sea dried up, there would have been little moisture available to produce rainfall over the land to the east of the Mediterranean Basin, and this series of events may have been responsible for finally cutting the link between African and Asian apes, with expanses of desert lying between their two homelands. (The link was also severed by the growth of great mountain ranges, including the Himalayas, as Africa continued to push northward and India too collided with Eurasia, beginning about 5 Myr.) The apes and other species of North Africa itself must also have been affected, with their woodland homes shrinking back into a heartland further to the south. It was just about at the time the Mediterranean Sea settled into its present-day form, around 5.3 Myr, that some ancestral form of ape that had lived through these convulsions, and whose own evolution had undoubtedly been affected by the accompanying climatic upheavals, gave rise to the three lines that would become ourselves, gorillas and chimpanzees. We can only guess at the exact nature of that ancestral ape, since no relevant fossils have ever been uncovered. But there is a wealth of evidence about how and where apes evolved into people over the next 5 million years, and about the climatic and other environmental changes that brought about this adaptation, driving our ancestors out of the woods and onto the plains. The previous chapter closed with the death of the dinosaurs, at the end of the Mesozoic era and the Cretaceous period; now, we have come to another logical place to break our story, not quite the death of all the apes, but certainly the end of the epoch of the apes, the Miocene. From now on, our tale is strictly that of the hominid line itself.

CHAPTER FIVE

Children of the Ice

Around 5 Myr, the world once again changed sufficiently for geologists to designate that time the end of an epoch, the Miocene. The new epoch that began at that time is called the Pliocene; for us, this is the most special of all past geological epochs. During the Pliocene, the first true hominids appeared on the evolutionary stage, and by the end of that epoch our ancestors were already sufficiently human to go by the name of *Homo*. Yet this was a much shorter interval than the Miocene, lasting only up until about 1.8 Myr. (Because it is easier to find and interpret evidence from more recent subdivisions of geological time, there is a tendency for the size of each division recognized by the experts to be smaller the closer we come to the present. This does not reflect an increased pace of geological or evolutionary change; it is simply that, just as when we view a broad landscape, we can see more detail close up.) On the traditional geological timescale the Pliocene was also the last epoch of a period, the Tertiary; in the new system of classification it is simply the second of four epochs in the Neogene. But either way it is, in a sense, our epoch.

Our line emerged during, and because of, another shift in the climate. But this change was much more subtle and long term than some of the dramatic changes, accompanied by mass extinctions, that mark the boundaries of other intervals of geological time. Indeed, one of the curious distinguishing features of the Pliocene is that it was an epoch when a *deterioration* in climate brought about what seems at first sight to be a *proliferation* of mammal species. To understand this apparent paradox, we have to stand back a little and look at the broad pattern of geological changes over the past 10 million years or so, and especially at the way the landmasses were slowly drifting into their present-day positions.

Drifting into drought

Little was changing in the deep south of our planet around this time, except that Antarctica was becoming increasingly isolated as both Australia and South America moved away from it. This may explain why the circumpolar current intensified around 6 Myr, producing the waves of extreme cold that locked up more ice than today in the Antarctic, and caused the repeated desiccation of the Mediterranean Basin. At that time, the layer of ice over the southern continent was several hundred metres thicker than it is now, with the sea level some 50 m lower than today and a desert in Austria as a result. But while Antarctica's increasing isolation was locking it into a pattern of ice, wind and ocean current that has persisted, with minor fluctuations, for many millions of years, the situation around the North Pole, where there had been open and relatively warm water since before the time of the dinosaurs, was changing.

As our attention focuses more finely on the evolution of the human line, we find that changes around the North Pole, not the South, come to dominate the story of the changing environment and its influence on our ancestors. Gradually, as continents moved northward into the Arctic zone, the climate in some regions changed. There is evidence of glaciation over the mountains of what is now Alaska as early as 10 Myr; this is an exception, linked with the growth of those mountains, and there is no sign of general northern glaciation until the end of the Pliocene itself. But it serves as a useful reminder both that conditions in the far north were getting colder, and that mountain-building plays a significant part in altering the climatic patterns of the world.

Around the time that the Mediterranean Sea was drying up, the Arctic was enjoying a cool, temperate climate, with coniferous forests extending to the northern limits of the land. Those limits were further north than they are today, because the fall in sea level that dried out the Mediterranean also left a broad expanse of continental shelf north of Alaska and Asia high and dry. By now, the continents were just about as far north as they are today, but this drying of the continental shelf pushed the boundary between land and sea several hundred kilometres northward of its present position. There may well have been seasonal snows at these high latitudes, but there is no sign in the geological record of widespread permafrost at that time, nor is there for another couple of million years.

One factor that may have helped the north to remain largely ice-free for a little longer was an intensification of the Gulf Stream, a warm ocean current which flows northwards up the eastern seaboard of North America, and carries heat to high latitudes. This is part of a circulation – a gyre – around the entire North Atlantic. The North Atlantic gyre itself may have intensified because of the changing geography of the globe, and especially as South America at last closed up on North America, with the gap between them being permanently bridged by about 3 Myr. Water that used to flow from the Atlantic to the Pacific between the two continents was increasingly diverted northwards into the Gulf Stream as the gap slowly closed after about 5 Myr.

About the time that the series of Mediterranean desiccations ended (partly because the Antarctic ice retreated slightly, but probably also, since the Sea has never dried out since, because of a geologically tiny widening of the gap between Spain and Africa), a new phase of mountain-building was again reshaping the face of the planet. We have already mentioned that the collision between India and Asia began to build the Himalayas to their present magnificence about 5 Myr; a little later the Andes in South America began to rise from a relatively modest height of 2000 m to their present 4000 m. As the Andes rose, they shielded the forests of the Amazon from rain-bearing winds (the prevailing westerlies), making the climate there drier. Similar changes took place in the lee of the Himalayas, increasing the geographical differences in climate. Meanwhile, in the northern continents the weather was becoming drier, but seasonal differences in climate were also increasing. The increasing strength of the pulse of northern seasons even extended its influence down into Africa, as the circulation of the atmosphere of the northern hemisphere began to shift rhythmically over the annual cycle.

Because there was a bigger variety of climates than there had been for tens of millions of years (at least), there was a greater variety of homes for life on the planet. When much of the land of the Earth was covered by what we think of as tropical jungle, then there was ample room for tropical species to live, but nowhere for any equivalent of modern temperate-zone species to evolve. It was tropical forests, or nothing. When the world became divided into tropical, temperate and other zones, that was bad news for the tropical species, who had their share of the real estate drastically reduced; but it meant that new species could evolve in different parts of the world as they moved out of the woods and adapted to fit the different conditions. So while the world became less of a tropical paradise for life, the variety of life on Earth actually increased. Modern dogs, for example, emerged about 6 Myr, with bears, camels and pigs following in a spurt of mammalian adaptation and radiation over the next couple of million years. At exactly that time, as the jungle of Africa shrank, there was a three-way split in the line of the African apes. It happened in East Africa, where the local variation on the mountain-building theme was lifting a great slab of the Earth's crust and cracking it apart to create a great rift-valley system. Instead of tropical jungle with a constant, all-year-round climate, the region became drier. The size of the rain forest was limited. While the amount of rain falling in any particular location depended on the new geography (on whether or not the region was in the lee of a new mountain), overall the climate became slightly seasonal, as part of the pattern of global climatic changes. 'Islands' of tropical jungle were now surrounded by grasslands and regions of more open woodland. And that, literally, is where we came in.

Tell-tale molecules

We know when the hominid line split from the line leading to other African apes because molecular biologists can now measure the differences between the genetic material, the DNA, of humans and other apes. (The technique is essentially the same as the genetic fingerprinting now used to resolve disputes about paternity, and has been used to identify rapists and other criminals.) This is part of the evidence which shows that *Ramapithecus* was not a hominid, but lived before the man–ape split. But some popularizations of the story of human evolution (and even some textbooks) fail to make the point that this does not necessarily mean that *Ramapithecus*, or at least one of the ramapithecines, could not have been an ancestor *both* of ourselves and of the other African apes.

We described in Chapter One the biochemical techniques by which the genetic 'distances', as they are known, are measured. The essential point to remember is that evolution proceeds because of the accumulation of changes in the DNA of species, changes in the genes themselves.* It happens that these changes happen at a more or less steady rate, at least in comparable molecules from closely related species. This is not something that you would necessarily expect to be true: people, for example, live longer than chimpanzees and have a greater gap between generations, so you might expect the molecular 'clock' to tick at different rates in the two species. But a battery of biochemical tests shows that this is not the case, and that the amount of change that accumulates in the DNA of human beings (their gene pool) over, say, a million years is roughly the same as the amount of change in the gene pool of chimpanzees in a million years.

These changes are all relative. They show, for example, that there is about a 1 per cent difference between the genetic material of human beings and either of the other two African apes, the chimp and gorilla. The 'distance' between ourselves and the orang-utan is, in molecular terms, about twice as great, showing that the line leading to both ourselves and the African apes split from the line leading to the orang-utan (and the other Asian apes) roughly twice as long ago as the three-way split in

* We are, of course, aware that some people are still uncomfortable with the idea of evolution, especially when we start talking about specifically human evolution. The best brief response we can make to those who feel such discomfort is to paraphrase what Stephen Jay Gould of Harvard University said in an interview published in *Newsweek* on 29 March 1982. Evolution is a fact, like apples falling out of trees (the fact is attested to by, among other things, the wealth of fossil evidence alluded to briefly in this book). Darwin's theory of natural selection was put forward to account for the fact of evolution, just as Newton's theory of gravity was put forward to account for the fact that apples fall out of trees. Newton's theory of gravity has now been superseded by a better theory of gravity, the one developed by Einstein. But apples did not stop falling out of trees when scientists began to debate whether Einstein's theory might be an improvement on Newton's. In the same way, some scientists now debate whether Darwin's theory of natural selection might not be improved upon; but species do not stop evolving because of that debate, and the willingness of scientists to debate the mechanism of the process does not indicate that they doubt the fact of evolution.

the African-ape line. If we can find just one accurate date from the fossil record, an accurate date for a split in the primate line, then all the other dates in the sequence will fall in to place.

The date which is used to provide this benchmark is the date of the split between Old World monkeys and apes, the arrival of *Aegyptopithecus* on the scene just after 30 Myr. The genetic distance between any of the living apes and Old World monkeys is about three times the distance between all the African apes and the orang-utan (perhaps a little more than three times this distance). So, if the distance between apes and monkeys corresponds to about 29 million years of separate evolution, the difference between African apes and the orang-utan must correspond to about 9 million years of evolution. By the same token, the difference between ourselves and the other two African apes corresponds to just half this, between 4 and 5 million years of independent evolution. The date for the human–ape split, inferred from the molecules in our blood, is smack in the middle of the time when the East African forests were fragmented and in retreat, with the region drying out and providing new habitats in the form of open woodland and grassy savannah. We can date the emergence of the hominid line to within about a million years, even though we have no fossil remains of our ancestors from exactly that time.

The big split between Asian apes and African apes occurred between 9 and 8 Myr, as the Miocene climate deteriorated and the forest linking the two groups disappeared. *Ramapithecus*, of course, was still around in both parts of the world at that time, and had been since at least as far back as 14 Myr. So there is no problem in accepting the fossil evidence that *Ramapithecus* was the ancestor of Asian apes such as the orang-utan. But did the African branch of the *Ramapithecus* family really die out? The molecules tell us that we share a common ancestor with Asian apes from about 10 Myr. The ancestor of the Asian apes that was around at 10 Myr was *Ramapithecus* (or at least a ramapithecine). So it seems entirely likely that our own ancestor from 10 Myr was also *Ramapithecus*. Indeed, it is difficult to come up with any other interpretation that matches both the fossil and the molecular evidence.

Ramapithecus disappears from the African scene at around 7 Myr, not that long before the date of the human–ape split. As the molecular technique has been refined in recent years, so the dates of this particular three-way split have been adjusted. Some researchers still argue that there was an almost simultaneous branching of some ancestral (ramapithecine?) line into lines leading to ourselves, chimps and gorillas. But other tests, using proteins from the tissue of living individuals, as well as their DNA, suggest that the gorilla line split off first, at around 8 Myr, and that the split between us and the chimps occurred a little later, at 5 Myr. The fit between the disappearance of *Ramapithecus* and the emergence of the gorilla line seems almost too good to be true, but is none the less impressive. Just maybe, it means that we really do have a more or less complete fossil record of our own ancestry all the way from the dryopithecines at 20 Myr down to the final split with the chimps at 5 Myr. We do not have fossils from between 5 and 4 Myr to confirm this interpretation of our ancestry. But the timing given by the molecules exactly matches the timing of upheavals caused by the changing climate. And we know *where* the hominid line began, because there are plenty of relevant fossils from the region of the East African rift just a little later, between 4 and 3 Myr. They show that the region had suddenly (by geological standards) filled with several varieties of 'ape-men' (in fact, one of the key discoveries is of the remains of an 'ape-*woman*'): our own ancestors and their very close relations. This radiation of the ape line must have been a response to the changing environmental conditions. But before we look at how those changing conditions drove our ancestors out of the woods, we ought to take stock of the almost bewildering variety of hominoid species that shared this corner of Africa at the time.

Entry of the hominids

The confusion is not helped by the way in which palaeoanthropologists keep arguing about the names that should be attached to different fossils, classifying and reclassifying species not quite at whim but in accordance with a set of rules sufficiently esoteric to leave most laypeople floundering. We shall try to stick to one set of names, without going into too much detail about the historical reasons why they were chosen. If you find different names attached to the same species in other books, do not worry too much about it. We shall also pick out, from several different ideas about just how these different species were related, the view that in our opinion provides the simplest explanation for the emergence both of the hominid line (leading to ourselves) and of the chimp and gorilla lines.

The earliest hominid that is given the genus name *Homo* was around in East Africa by 2.5 Myr. Dubbed *Homo habilis*, this was an ape that walked upright (standing about 1.2 m tall), had a fairly slight build and a large head with a brain averaging 675 cm^3 in size, just half that of a modern *H. Sapiens*. By 1.5 Myr, this had developed into *H. erectus*. Taller (1.6 m) and with a larger brain (925 cm^3), this was the species that spread the *Homo* line out of Africa and into Asia. Both species were tool-users. By 0.5 Myr *H. erectus* had evolved into *H. sapiens*, the modern human form that spread around the world. But the story of *Homo* is for the rest of this book.

Homo habilis shared the region of the African rift valley with two very close relations, members of a genus called *Australopithecus*. *Australopithecus africanus* was the first on the scene, and fossils of this species have been found in strata 3.5 million years old. The youngest-known fossils of *A. africanus* are dated at 1 Myr. They too were upright walkers with a slender build, about the same size as *H. habilis*, but with a smaller brain, only 440 cm^3 in volume (not much different from the average for modern gorillas, about 500 cm^3). There is some evidence that *A. africanus* were tool-users, but if so they were unsophisticated compared even with the stone tools used by *H. habilis*. The other species of *Australopithecus* contemporary with *H. habilis* was bigger and more heavily built, and is known as *A. robustus*. They were around between 2 and 1 Myr, but no younger fossil remains have been found. Some members of *A. robustus* stood 1.6 m tall, and their average brain size was a little bigger than that of the average modern gorilla, about 520 cm^3 (the biggest gorilla brains, however, check in at around 700 cm^3). But there is no evidence that they used tools.

Both the *Homo* and *Australopithecus* lines had emerged by 3 Myr, clearly descended from a recent common ancestor, but equally clearly two separate branches of the evolutionary tree. Where did all these creatures come from? And how were they related? The debate on those issues raged for decades among the experts; indeed, it still continues. But the answers may have been found in Ethiopia in the second half of the 1970s. There, Donald Johanson and colleagues discovered, over a lengthy period of fossil-hunting, the most complete skeleton of any human ancestor more than 35,000 years old. Informally known as Lucy (after the Beatles song 'Lucy in the Sky with Diamonds'), the skeleton is 40 per cent complete, the remains of a female ape-person from about 3.3 Myr. She belongs to a species that was given the formal name *Australopithecus afarensis* in 1978, after much heart-searching by the discoverers, who would dearly have liked to place her in the genus *Homo*. It remains a moot point whether *A. afarensis* or *H. afarensis* is the better name, but the simplest interpretation of the available evidence is that the species Lucy belonged to was ancestral to both *Homo* and *Australopithecus*. She walked, the structure of her skeleton shows, if not completely upright then certainly better than a chimpanzee walks today, and she stood just over 1 m tall (other members of what may be the same species from the same size were as much as 1.7 m tall, suggesting to some researchers that the males towered over the females), but she had a small brain; the best instant image of Lucy is of a proto-human body with an ape's head on top of it. Walking was, it seems, quite literally the first step on the road to being human; the big brain that we are so proud of came later, perhaps because of the opportunities provided by upright walking to free hands for complex tasks (such as tool-making) that needed a good brain.

Astonishingly, at about the same time that Johanson's team was uncovering the remains of Lucy in Ethiopia, a thousand miles away in East Africa Mary Leakey discovered the actual fossil footprints made by a similar creature walking upright over a layer of volcanic ash that formed about 3.7 Myr. They were preserved by a remarkable series of coincidences. First, a volcanic eruption spread a layer of ash rich in carbonatite, a kind of natural concrete that sets solid if it is first wetted and then dried. Then, a shower of rain dampened the ash, and while it was still wet what Roger Lewin calls 'a veritable menagerie' of creatures hurried over it, perhaps fleeing from the eruption. Hares, baboons, antelopes, two types of giraffe and a kind of elephant were among at least twenty different species that left their marks in the 'concrete' before it set, along with two hominids joining in the exodus. The hardening layer of ash was then covered by more ash and by windblown dust and soil, remaining covered up until weathering exposed it to view in 1976. The hominid footprints that were revealed must have been left by very close relatives, anthropologically speaking, of Lucy. They were made by a large hominid and a small one walking side by side, and in the light of Johanson's discoveries it is tempting to guess that they were a male and female of the same species. *A. afarensis*. But this is only a guess.

Indeed, there is still a great deal of guesswork and argument about the significance of all these finds. We do not want to go into all the details; Roger Lewin has already done that, in *Bones of Contention*. What we are interested in here is that both the fossil remains of Lucy herself and the presence of these fossilized footprints show that our ancestors set out on the path to becoming human by learning (or being forced) to walk upright. At the same time that some of the descendants of *A. afarensis* were improving their walking skills and developing bigger brains, two other descendants, *A. robustus* and *A africanus*, seem to have died out. At least, that is the conventional wisdom. But there is at least the possibility, which we discussed in *The One Per Cent Advantage*, that these two australopithecine lines actually survived and evolved to become modern gorillas and chimpanzees. In order to make the pieces of the puzzle fit together in this way, you would either have to adjust the molecular timescale a little (which is possible, but not easy), or you have to accept that the split between our line and the two australopithecine lines contemporary with *H. habilis* occurred rather before Lucy's day, so that she is really a member of *H. afarensis*. The second possibility seems quite likely, in view of the molecular evidence that the human–ape split was at around 5 Myr, but the puzzle then is why there are no fossils of *A. robustus* or *A. africanus* from around Lucy's time. In any case, at present we are stuck with the official name *Australopithecus afarensis*, and if that is what the experts have decided, it seems logical to place the split (with some slight misgivings, and acknowledging that other interpretations are possible) after Lucy. Fortunately, this is very much a side issue from the story we have to tell here. Whatever happened to the australopithecines, and whatever formal name you give Lucy, we can now see, from the evidence of climatic upheaval in East Africa around 4 Myr, just why our ancestors gave up the life of a tree-dwelling primate and tried their luck out on the plains. Putting it at its simplest, it was not so much that Lucy and her like left the forests, but that the forests left them.

The bipedal brachiator

The East African rift-valley system actually runs all the way from southern Turkey to the mouth of the Zambezi river, through Israel, the Red Sea and the lakes of East Africa itself. In places, the valleys associated with the system are 80 km wide and 300 m deep; and it is laced with active volcanoes. The system is the growing product of the same kind of plate-tectonic activity that smashes continents together to throw up great mountain ranges – or, rather, the opposite kind of activity. Along the jagged line of the Great Rift Valley, the Earth's crust is being torn apart by sideways forces associated with a long, slow turnover of fluid material in the hot depths below. These are the same kinds of force that tore apart Laurasia and Gondwanaland, creating the present continents and setting in motion the present spate of continental drift. A hundred million years from now there may be a great ocean separating the bulk of Africa from the horn of Ethiopia and Arabia, newly formed sea bed whose growth matches the destruction of old sea bed at other sites around the world, especially along the western rim of the Pacific. But for the present, and in the immediate geological past which is of special interest to the story of human origins, the most important feature of the African part of the rift-valley system is that it provides an enormous variety of habitats. There are dramatic differences over a short geographical range, providing an environmental patchwork, because of changes in the altitude and in rainfall and water-drainage patterns over short distances. Instead of a uniform spread of tropical jungle, patches of dense forest are interspersed with more open woodlands and grassy savannah.

How does a tree-dwelling primate adapt when the forest in

which it lives begins to shrink? There are two obvious solutions. Either you retreat into the heart of the forest, and carry on your business of finding food to eat and a safe place to sleep as before, or you step out into the new surroundings and try to make a living in a new way that suits the changing world. Without the individuals concerned making any conscious choice along these lines, this is how evolutionary pressures work to select new varieties from a common ancestral stock at times of environmental change.

There is no need, we are sure, to labour the point; you see what we are driving at. When the forests shrank, some apes became more 'ape-like', sticking with the trees and, if anything, becoming more efficient at living there in the face of increasing competition for dwindling resources. Their descendants are the gorillas. Other apes, originally members of the same species, scrambled a living in the more open woodlands nearby, still among trees (although not thick forest), but forced to find food on the ground as well. As evolution fitted them better for this lifestyle, they became chimps. And a third branch of the family, perhaps descended from the individuals who were *least* adept at the old lifestyle and were pushed out onto the plains by competition from their cousins, had to find a completely new way to live, or die. They developed upright walking, were forced to eat almost anything that came to hand, and learned the value of sharing their food amongst a family or larger group. Eventually, they developed large brains, and became human.

All this must have been a gradual process. Stating it so baldly makes it sound like an overnight transformation: with one bound, the primitive ape leaped out of the forest, stood upright and became human. In fact, the adaptation was gradual, unnoticeable from one generation to the next but building up over hundreds of thousands of generations, and millions of years of time. The apes that 'stayed apes' need not actually have retreated into the shrinking jungle. They were probably the ones that happened to live in the bits of jungle that were still there after the environment changed (and remember, this time the climate changed *slowly*). The apes that 'became human' did not physically move out of the jungle, but over a long period of time the climate in the valleys where particular groups of apes lived dried out, and the trees slowly disappeared. In each generation, individuals that learned to cope with the changing conditions a little better than their contemporaries would be the ones that found most food and had the best chance of rearing offspring. If upright walking, for example, made it easier to find food and carry it back to the family, then natural selection would favour upright walkers, and the characteristic bipedalism of the human ape would evolve.

One interesting piece of corroborative evidence that it was the changing climate that made humans out of apes comes from the monkey line, from which we split almost as far back as 30 Myr. Since the heyday of the apes in the first half of the Miocene, monkeys have been far more successful than we apes, in terms of the number of species around on Earth, and the handful of ape species today is greatly outnumbered by more varieties of monkey than you can shake the proverbial stick at. But monkeys, too, are primarily tree dwellers, and when the forests declined at the end of the Miocene, in some regions monkeys also had to adapt or die. At just about the same time that the hominid, chimp and gorilla lines were getting started, between 5 and 4 Myr, the modern baboon line split from an ancestral monkey stock. Baboons are monkeys that have adapted to life on the open plains, and use trees only for refuge. Humans and chimps represent the apes' 'answer' to the problem of shrinking Pliocene forests; baboons represent the monkeys' 'answer' to the same problem. But baboons, unlike people (and unlike Lucy), cannot walk upright. It may be that the world today is dominated by an ape-descended species, rather than monkey descendants, for one main reason: the fact that the lifestyle of a tree-dwelling ape gives it a body structure that, coincidentally, contains many of the features appropriate for standing and walking upright.

The point is that, whereas monkeys are relatively small, light creatures that run along the tops of branches, and leap from tree to tree, apes are relatively large, heavy creatures that swing hand over hand under the branches, and also swing, rather than jump, from one branch to the next. This means of locomotion is called brachiation. Of course, monkeys can hang from a branch by their hands (although they never move along the branch while dangling in this way), and an ape can scramble on top of a branch, if it finds one sturdy enough. But the habitual lifestyles of the two primates are different, and the structure of their bodies reflects evolutionary adaptation to these different means of locomotion. It only takes a moment's thought to appreciate that brachiation and, literally, hanging around in trees favours the evolution of an upright body structure. Whether hanging from your own two hands or standing on your own two feet, the basic position of your body is the same. Our brachiating ancestry is clear from many anatomical features of human beings, features which we do not share with monkeys. For example, you can bend your wrist in a right angle in the direction of your little finger, an obviously useful degree of flexibility when swinging through the trees. And you have (potentially) strong biceps, that, if you bother to keep in training, can lift you up onto a bar fixed overhead. Human shoulder-joints allow far more freedom of movement than their monkey counterparts. And so on. Try kneeling on all fours, and you will find that the only view you get, unless you force your head back painfully, is of the ground. Standing upright, your head is ideally placed for looking around; but the position of the head on top of the spinal column is also just right to give a good field of view to an ape hanging by its arms from a branch. So, we are descended from brachiator stock, and the anatomical adaptations required by brachiation and evolved over millions of years in the Miocene already suited the ape body, in many ways, to the task of standing upright. Lucy was such a good walker that her ancestors (possibly ramapithecines, or ramapithecine stock) must themselves have already begun to move out of the woods and onto the plains. We have seen why they, and their descendants, had to make a go of a new lifestyle in a changing world. But how did their rapidly developing ability to walk upright make them successful in that new environment, a couple of million years *before* they evolved into big-brained *H. habilis*?

An upright stance offers many advantages out on the plains. For a start, it gives you the best possible view of your surroundings, both to find food and to get advance warning of predators coming your way. Secondly, once bipedal locomotion has evolved, it proves a more efficient means of getting around on the ground than the scrambling gait of the chimpanzee, and uses less energy. But, most important of all, many palaeoanthropologists agree, it frees the hands to carry things. The things might be weapons, even if only stones that are thrown to bring down prey or scare off attackers; but perhaps the best explanation of the success of the upright lifestyle is that it enabled our ancestors to carry a variety of different kinds of food, gathered here and there, back to the family group to share. There is archaeological evidence from sites in East Africa 2 million years old that by then our ancestors were spending a lot of time at certain locations. They left behind them discarded stone tools and bones from many different animals – animals which certainly did not just wander into the camp and drop dead, but whose bones, still covered in meat, were carried there by people. Those lumps of meat, however, may have been a minor part of the group's diet. Unfortunately, any seeds and food plants that were brought into the camp for consumption have left no fossil remains; but there are other ways to find out about this kind of lifestyle.

Tribes that lived in much the same way, hunting for some of their food and gathering plants to share as well at a settled site, survived in some parts of the world until well into the twentieth century, and their lifestyle has been studied by anthropologists for clues as to how our ancestors lived. There is a typical division of labour between the sexes in such hunter-gatherer societies, with males sent out to hunt while females look after the children and gather plants to eat. Lest we be accused of rampant sexism in pointing out this basic fact of human life, we should also emphasize that the males are not necessarily very good providers: the group usually depends on the women and the food plants they gather for survival, with the occasional contribution of meat from the men a welcome bonus. Only slightly facetiously, some observers quip that the main reason for the men being sent out hunting is to keep them out of the way while the women get on with the real work. But this tongue-in-cheek comment conceals the important truth that, by making different individual efforts to find different kinds of food, the group as a whole (and the individuals who are part of the group) has more chance of survival because each individual is not dependent on a single source of food that might, for some reason, become unavailable one particular day or during one season. As for the hunting itself, this certainly began merely as scavenging, picking up leftovers from the kills made by the really efficient hunters, such as the big cats.

Early humans may not have been very good at anything, except walking upright; but they managed, with the aid of their upright stance, to make a reasonable job of doing each of a lot of different things. They became *un*specialists. But whatever the exact sources of all the different food stuffs needed to make the unspecialist, food-sharing lifestyle work, the key requirement is the ability to carry things back to camp to share. You need hands to carry things, and that only leaves you with two limbs free for walking.

Before even becoming intelligent, the human line became a bipedal brachiator. The change to bipedalism happened in response to changing climate and environmental conditions in the rift-valley system of East Africa; but for several million years bipedalism gave the erstwhile brachiators a successful lifestyle, without the need for intelligence. What was it that forced the bipedal brachiator to become intelligent, after about 2 Myr? Almost certainly, another turn of the climatic screw, or rather, several turns. By 3 Myr, the ice had begun to spread in the north. At first, the onset of the northern ice age had little impact in East Africa. But its influence began to be felt more strongly, and in an unusual way, just at the time that *H. habilis* was emerging. Around the end of the Pliocene and the beginning of the Pleistocene, 1.8 Myr (at the beginning of the Quaternary period on the old time-scale), there was a dramatic decrease in the rainfall in the region, linked with the spread of ice at high latitudes. This heralded the onset of a series of ice-age rhythms that brought unique environmental pressures to bear on the bipedal brachiators. The children of the ice were about to appear on the African scene.

Had it not been for climate-related environmental changes, we would still be Miocene apes living in the trees. The first phase of cooling in the northern hemisphere and drying in Africa was sufficient to push our ancestors out of the woods and make them upright walkers. But the series of climatic changes that put a premium on intelligence, and set us on the road to being human, were much more subtle and complicated than a mere cooling of the globe. They turned an ape (albeit an upright one) into *H. sapiens* in less than 3 million years, a breathtakingly fast spurt of evolutionary change. This happened in response to a climatic situation that is extremely rare, and may even be unique, in the long history of our planet. But it was only in the past couple of hundred years that the intelligent ape realized that the climate of the Earth had recently been through a series of convulsions, now known as an ice epoch; and it was only in the present century that we first appreciated that this ice epoch began a couple of million years ago.

The idea that great glaciers had once covered large parts of Europe that are now ice-free occurred to several people in the late nineteenth century. Before that time, observers of the natural world had already noticed that so-called erratic boulders are sometimes found in places where they do not belong, far from any rock formations that they match, and that the landscape of parts of Europe is dotted with jumbled heaps of rocks and sediment. To most seventeenth-century geologists, however, it was 'obvious' that this was the work of the biblical Flood, washing all before it. So it is ironic that in 1787 one of the first people to make the case that great sheets of ice, not a flood of water, were responsible was a Swiss clergymen, Bernhard Kuhn. Other scientists came up with similar ideas independently, including the Scot James Hutton, who visited the Jura Mountains of France and Switzerland in the 1790s; he was impressed by the scars caused by glacial activity that he saw in the rocks there. Hutton's espousal of the idea of an ice age also contains a minor irony. It was Hutton, after all, who first promoted the idea that the surface of the Earth had been shaped not by biblical catastrophes, such as the Flood, but by the same processes that we can observe at work today, operating slowly over long periods of geological time.

This idea, the principle of uniformitarianism, implied a much greater span of Earth history than the timescale inferred from a literal interpretation of the Bible, and stirred controversy in the late eighteenth and early nineteenth centuries. Today, a modified form of uniformitarianism, and a long history of the Earth, are cornerstones of geology; and yet, as we have seen, the Earth does also suffer catastrophes, on the appropriate timescales. And one of those catastrophes is the ice epoch whose importance Hutton was one of the first to appreciate. Of course, there is no real conflict with his ideas of uniformitarianism, since the key phrase is 'on the appropriate timescale'. On a long enough timescale, catastrophe is 'normal'; in terms of the long history of the Earth, an ice epoch is one of the natural processes of change that shape the planet, like the action of volcanic eruptions (also pretty catastrophic, if they happen alongside you), or erosion by running water and tides. But neither Kuhn nor Hutton, nor anyone else of their generation, made much effort to persuade the scientific world of the time that there really had been an ice age. It was only in 1837 that Louis Agassiz, the 30-year-old President of the Swiss Society of Natural Sciences, took up the case and promoted it with such vigour, in the face of almost equally vigorous opposition from biblical catastrophists, that by the middle of the 1860s the ice-age theory had become widely accepted.

Evidence of ice

The evidence that persuaded Agassiz that ice had scoured the European landscape was all about him in his native Switzerland. But his eyes were opened to the meaning of the marks of glaciers on the mountains by the work of Johann von Charpentier, a pioneering glaciologist born in Freiburg, Germany, in 1786. Carpentier moved to Switzerland, later adopting the French version of his name, Jean de Charpentier (and later still, in 1855, he died there). He worked as a mining engineer, becoming director of salt mines in the canton of Vaud. He was intrigued by glaciers and impressed by their power. Working sometimes with other amateur geologists (including a civil engineer, Ignatz Venetz-Sitten, who introduced him to the puzzle), Charpentier made a special study of the locations of large boulders which seemed to be made of rock from the Swiss Alps, but are found today far away down the Rhône valley. By 1834, he had reached the conclusion, like others before him, that these huge, 'immovable' boulders could have got there only by being carried in the grip of great glaciers that had long ago slid down from the mountains during an ice age. He presented his evidence that year to a meeting of the Society of

Natural Sciences, but seems to have persuaded nobody, not even Agassiz.

Indeed, when young Agassiz heard of Charpentier's work, not only did he not believe a word of it, but he set out to prove that the whole idea was nonsense. He was well placed to do so, since in 1832 he had been appointed Professor of Natural History at the University of Neuchâtel, and he was becoming known as an expert on fossil fishes. As a native of Switzerland, he was sure from everyday experience that glaciers could not move far enough, or fast enough, to transport great boulders down into the Rhône valley. As a good scientist, he set out to prove his case. He set up an observing station in a hut on the Aar Glacier, and carefully measured the movement of the glacier (and others) by driving stakes into the ice and measuring their movement. Confounded by the discovery that the ice moved much faster than he had thought, and persuaded that it could also carry large boulders along with it, Agassiz was converted from sceptic to believer, and like many converts he became an enthusiastic evangelist for his new beliefs.

He started, in 1837, by dragging his reluctant fellow members of the Society of Natural Sciences out of the lecture room and into the mountains to see the evidence for themselves. They were not immediately convinced, and some even preferred to argue that marks in the rocks might have been made by the wheels of passing carriages, not by the grinding of rocks carried by glaciers. But Agassiz was not to be dissuaded. He went out with Venetz and Charpentier to look at the evidence farther afield, but soon raced ahead of them in his enthusiasm, propounding a wide-ranging theory of a world covered by ice, which he published in 1840. Charpentier himself, unable to ride the whirlwind he had helped to generate (and more than a little put out by the way Agassiz had run off with the ball), published his own version only in 1841. But even if he had beaten Agassiz into print, there is no doubt which version of the story would have made more impact. When Agassiz wrote about ice ages, nobody else could compete with him for dramatic imagery, which gives even the tabloid journalists of today a run for their money:

The development of these huge ice sheets must have led to the

> destruction of all organic life at the Earth's surface. The ground of Europe, previously covered with tropical vegetation and inhabited by herds of giant elephants, enormous hippopotami, and gigantic carnivora became suddenly buried under a vast expanse of ice covering plains, lakes, seas and plateaus alike. The silence of death followed ... springs dried up, streams ceased to flow, and sunrays rising over that frozen shore ... were met only by the whistling of northern winds and the rumbling of the crevasses as they opened across the surface of that huge ocean of ice.

They do not, alas, write scientific papers like that any more (indeed, few scientists did in the 1840s). Although Agassiz did get just a little carried away, and (as we shall see) life did not completely disappear from the face of the Earth (or even Europe) during the ice ages, it was that knack for imagery and publicity that helped Agassiz to convince his colleagues that the evidence for ice ages had to be taken seriously. But he would have had no case to make, of course, if painstaking pioneers like Charpentier had not tracked down and studied the huge erratic boulders. And even then, it took the best part of thirty years, and a lot more evidence, to persuade the doubters.

The evidence came in from around the world. In 1852, the awesome extent of the Greenland ice cap was mapped for the first time; later in the nineteenth century the size of the Antarctic ice sheets became known. And meanwhile, more traces of the activity of glaciers long gone were found not just in Europe but also in North America. We know now that a ridge of glacial debris, 50 m high in some places and marking the southern limit of the latest advance of the ice, runs all the way from Long Island in the east to Washington State in the west. In 1846, Agassiz himself visited the USA to study such remains of glacial activity; he stayed on as Professor of Zoology at Harvard University, married Elizabeth Cabot Carey in 1850, and became a pillar of American science until his death in 1873 at the age of 66.

Ice over the Earth

In the second half of the nineteenth century, and into the twentieth, the full extent of the great ice sheets of long ago gradually became clear. Over a period of several million years, huge areas of land around the north polar region of the globe have been scoured by ice sheets. But there was never a single, continuous sheet of ice uniformly distributed around the pole. The Arctic Ocean was covered by a skin of ice, just as it is today, and the southern limit of this ice reached much farther into the Atlantic than at present, so that both Greenland and Iceland were set in a frozen sea surrounding them on all sides and stretching away to the south. Both these islands, indeed, were completely covered by their own glaciers – more than completely covered, in a sense, because as the sea level fell (because water was locked up as ice) there was more dry land around the islands, and at the edge of continents, on which the ice sheets could rest.

The glaciers of the Alps, which left the traces that led the pioneers to the realization that there had been an ice age, were ironically, relatively small beer by ice-age standards, isolated (but large) mountain glaciers lying well to the south of the Eurasian part of the northern ice cap. This region of glaciation, known as the Scandinavian ice sheet, covered about 6.6 million km^2 over present-day Europe, from Britain in the west across the Baltic to modern Russia. It also pushed northwards over the Arctic sea bed to link up with glaciers in Spitzbergen, adding another 500,000 km^2 to its area, and linked with Siberian glaciers to the east. But in the most easterly parts of what is now Siberia, there was no ice cap, because the world was too dry; and no moisture-laden winds could penetrate there to dump their burden of snow.

The dry region continued across a land bridge between Asia and North America, where the Bering Strait had dried out as the sea level fell. Most of Alaska was free from ice. But the greatest of all the ice-age glaciations (the Laurentide ice sheet) covered a large part of North America, including all of present-day Canada. The southern limit of the ice ran roughly from the site of New York City to the Rocky Mountains in Montana, and it covered an area of more than 13 million km^2, bigger, on its own, than the Antarctic ice cap today (although the ice was not as thick as the Antarctic ice is). The Rockies themselves poked up through the ice, but another glacier system ran between them and the Pacific, 2.3 million km^2 of ice in a ribbon from Alaska to Washington, Idaho and Montana. The southern edge of the ice ran roughly through modern Cincinnati and St Louis, past Kansas City and across St Pierre, South Dakota. On the other side of the world, ice covered the locations of Dublin, London, Amsterdam, Berlin, Warsaw, Kiev, Moscow and St Petersburg.

Ice cover also expanded in the southern hemisphere, although less extensively. Antarctica itself, as we have seen, had long since been in the grip of ice, and could hardly become more glaciated. But mountain glaciers have spread considerably, both in South America and in New Zealand (especially the South Island) at various times during the past few million years. Many smaller glaciers gripped regions such as Tasmania and Japan, and spread over the mountains of China, as well as the mountainous regions of Europe and the Rockies south of the Laurentide ice sheet.

In North America and Eurasia south of the ice, the climate was both cold and dry. But, like the increased glaciation in the southern hemisphere since the end of the Pliocene, this is of only peripheral interest to our story. The main point is that when the northern ice advanced, the weather also became dry (and a little cooler) in East Africa. There, climatic changes associated with the growth of great ice sheets to the north had a direct effect on our ancestors, the bipedal brachiators. In order to relate events in East Africa with climatic changes further north, it is crucial to have a timescale of the ice epoch, not only so we can relate the beginning of the ice epoch with the changes taking place further south which helped to drive our ancestors out of the woods, but also because the ice did not just arrive one day some time around 2.5 Myr and sit there quietly until about 10,000 years ago. Instead, it advanced and retreated in rhythmic waves of glaciation, producing a long, slow pulse of climatic change that put intelligence at a premium and made us human.

Ice-age rhythms

The Pliocene ended, and the Pleistocene epoch began, when ice first spread to cover a large part of the northern hemisphere. Nobody can put a precise date to that event, but we can say why it happened. It happened because the jostling of the continents in their drift around the globe and to higher latitudes finally sealed off enough of the flow of warm water into the Arctic Ocean for the skin of that ocean to freeze. Once it froze, it reflected away incoming solar heat in the summer, and caused a severe chill to spread across the land nearby.

Although there are traces of Alaskan glaciation even from Miocene times, there was no significant advance of the ice in the north until the late Pliocene, a little before 3 Myr. From this time there are signs of glacial activity in Iceland, and other evidence of a build-up of ice. By 3 Myr the Arctic region may have been nearly as cold as it is today, and although the snow and ice had still to spread farther afield, a pronounced cooling and drying began in equatorial regions about this time.

Ice first appeared on the continent of Europe around 2.5 Myr, about the same time that local glaciers formed on the mountains of California. There is no direct evidence available of the onset of glaciation in Greenland, or rather, if there is any evidence it is all buried under the ice cap today. But it seems likely that the glaciers began to grow there about the same time they began to grow in nearby Iceland. Through all this period – from 5 to 2 Myr – the Antarctic ice cap was bigger than it is today, and sea ice, especially the Ross Ice Shelf, extended farther out from the continent. But temperatures in Europe were still a little above those of today, and because the mountains were not so high as they are now, rainfall penetrated more effectively into the heart of the continent than it now can, encouraging the spread of forests. Some time between 3 and 2 Myr, however, as the fall in the sea level shows, the amount of water locked up as ice in the Greenland and Antarctic ice caps combined was more than the amount of ice in the polar regions today. At present, the area covered by ice is about 15 million km^2. At the height of Pleistocene glaciation the area covered was 45 million km^2, and the volume of ice reached 56 million km^3. Some people define the start of the present ice epoch as the time when the ice cover first exceeded that of the present day. Geologists, however, prefer to date the beginning of the Pleistocene (which for most practical purposes is synonymous with the ice epoch) from a convenient reversal of the Earth's magnetic field at 1.8 Myr which left its mark in rocks around the globe.

By a happy coincidence, this is almost exactly the time of the earliest-known fossil remains of *H. erectus*, the first member of the *Homo* line to move out of Africa. In terms of the evolution of our own line, the arrival of *erectus* on the scene would in itself make a good excuse to date the start of the Pleistocene to just after 2 Myr. Distinctly more human than its predecessor, *H. habilis*, a typical early member of *erectus*, had a brain with a volume between 800 and 900 cm^3, and some later specimens have cranial capacities as great as 1100 cm^3, getting close to the present-day human average of 1360 cm^3. *Erectus* had more modern teeth, and a less prominent jaw than *habilis*, and 'below the neck', says Roger Lewin, '*Homo erectus* was essentially human, except in the substantial robusticity of the limbs and muscle attachment points and in having a slightly shorter stature'. It was a fully erect (hence the name), upright biped which spread throughout Africa, Asia and Europe.

The *Homo* line emerged during the early phases of northern-hemisphere cooling, before the ice had really taken a grip on Europe and North America. But the glaciers began to spread at mid-latitudes by about 2 Myr, just before the somewhat arbitrary date of the beginning of the Pleistocene. *Homo habilis*, the ancestor of *erectus* and the descendant of Lucy, was on the scene then, and had, in a sense, been created by the changing climate which had brought the spread of conditions unsuited to woodland apes. But just as the early phase of cooling and drying of climate was only a transition from an ice-free northern hemisphere to a state of full glaciation, so *H. habilis* was only a transitional species from the australopithecines (including Lucy) to *Homo*. It takes time for natural selection to produce an evolutionary response to a change in the environment, and, hardly surprisingly, the changes in the *Homo* line lag a little behind the major changes in climate, but only by a little.

Remains of *habilis* are found from about 2 to 1.6 Myr. The line changed so quickly, producing *erectus* from the *habilis* stock, because the climate continued to change, and because after about 1.8 Myr it was changing in a different way. For at least 3 million years there had been a slow deterioration in the climate of the northern hemisphere, a slide into an ice epoch that made its influence felt farther south as a progressive drying of East Africa. But about the time that *H. habilis* made its brief appearance on the evolutionary stage, the first wave of full-blown northern glaciation had given way to a new climatic rhythm in which the glaciers advanced and retreated many times. During the Pleistocene, the northern ice repeatedly expanded to the point where roughly 30 per cent of the entire land surface of the Earth (including Antarctica) was covered by a blanket of ice, and just as many times the northern ice retreated into its Arctic fastness, where it is today. This regular rhythm has only been fully understood since the middle of the 1970s, partly because, for more than a hundred years after Agassiz finally persuaded his colleagues that there had been 'an ice age', nobody had fully appreciated just how many separate ice ages there have been during the ice epoch of the past 1.8 million years.

At first it was difficult enough for the scientific community to accept that there had been one ice age. Then, as geologists began to examine the heaps of rock and other traces of glaciation in Europe and America, they realized that the ice had advanced not once but several times. By the early twentieth century, it was generally agreed that there had been four major ice ages in the past half-million years or so, each lasting for perhaps a few thousand years, and separated by much longer intervals of warmth. Since the most recent ice age ended around 10,000 years ago, this was a fairly comforting picture. But the development of better techniques for dating geological remains and new techniques for determining the temperature of the Earth in years long gone eventually overturned this cosy picture. It now seems that another ice age may, by geological standards, be imminent.

One of the key developments that led to this realization is a technique known as the 'isotope thermometer'. It depends on the fact that atoms of common elements in the environment, such as oxygen and hydrogen (which together make up molecules of water) come in different varieties, known as isotopes. Take oxygen as an example. Most of the oxygen in the air that we breathe and the waters that cover the world is in the form of atoms of oxygen-16, where the number indicates their atomic weight. A small but significant minority are in a form known as oxygen-18, chemically identical but two units heavier (one unit, on this scale, is the weight of an atom of hydrogen, the lightest element). A molecule of water that contains an atom of oxygen-18, instead of oxygen-16, is correspondingly heavier and will find it harder to evaporate from the ocean than its lighter counterparts, but it will condense more easily in rain or snow. At the same time, tiny creatures that live in the sea (planktonic foraminifera) use oxygen from their surroundings, among other things, to build the chalky calcium carbonate of their skeletons. Because the temperature of the water in which they live affects their metabolisms, the proportion of oxygen-18 that these creatures absorb depends on the temperature of the water. When the creatures die, their calcium-carbonate skeletons sink to the bottom of the sea, building up layers of chalk. By drilling long cores from the sea bed and extracting samples of different ages (which means from different layers) for analysis, geologists can use the varying amount of oxygen-18 in the sediments as an isotope thermometer that reveals how the temperature of the upper layers of the ocean, in which the animals that made those skeletons lived, has varied.

All this, of course, is very far from easy. Even extracting long cores of chalky sediment from the sea bed requires sophisticated drilling technology, and analysing the cores by measuring isotope ratios is a painstaking business – not to mention the care needed in dating the cores, with the aid of their fossil magnetism and other geological markers. Although the idea was first proposed in 1947, by Harold Urey of the University of Chicago, it was not until the mid-1970s that a reasonably accurate chronology of recent ice ages was established (the full story up to that point is told by John Imbrie and Katherine Palmer Imbrie in their book *Ice Ages: Solving the Mystery*). Since then the technique has been developed further, confirming and extending the picture that emerged in the 1970s. The technique will always be a little imprecise, in terms of both the exact temperatures and the exact dates it yields, but it can certainly tell the difference between an ice age and the warmth of the world today. Instead of four ice ages separated by much longer intervals of such warmth, we are now sure that there were six ice ages in the past few hundred thousand years, and well over a dozen full ice ages during the past 2 million years. Not only were there more ice ages than used to be thought, but each one lasted longer. Each full ice age, with ice and snow covering 30 per cent of the land surface of the globe, lasts for about 100,000 years (this is a rough figure; some ice ages are 'only' 70,000 or 80,000 years long); the intervals of warmth between ice ages, like the conditions in which we live today, are called interglacials, and last for only about 10,000 years (this figure is equally rough; some interglacials may last as long as 15,000 years). We live in an interglacial that began about 15,000 years ago and in which the ice had fully retreated from mid-latitudes by 10,000 years ago. This is the basis for gloomy forecasts that the next ice age is due; the occurrence of so much climatic variation over a period of 2 million years or so is why the whole long interval of cold is best referred to as an ice epoch, although the glaciers have advanced and retreated many times within that interval.

The idea that there had been 'an ice age' became widely accepted only in the 1860s, and the idea that there had been so many ice ages became accepted only in the 1970s, more than a century later. But it did not take anything like as long to explain the new discoveries; indeed, in a sense it took no time at all. The discovery of the climatic rhythm of the past million years or more needed no new theory to explain it in the 1970s, because a detailed explanation of why climate should vary with a rhythm 100,000 years long, with short interglacials dividing much longer ice ages, already existed. It appeared in its first detailed form in the work of James Croll, a Scottish thinker who published his first paper on the subject in 1864, barely as the ice-age theory itself was achieving respectability. His work on ice ages, and that of a later pioneer, Milutin Milankovich (who developed the idea even further, and lent it his name) was to stay out in the cold for as long as 'everybody knew' that there had been only four ice ages in the past half million years, and that ice ages were much shorter than the intervals of warmth that separated them. When what everybody knew was overturned by the isotope thermometer in the 1970s, the theory of Croll and Milankovich was ready to come in from the cold.

The theory that came in from the cold

Like Agassiz, Croll followed in the footsteps of earlier pioneers. The idea that changes in the Earth's orbit as it moves around the sun might influence our climate can be traced back to Johannes Kepler, a seventeenth-century astronomer. But it was Croll, from the 1860s, who produced the first fully developed version of the astronomical theory of ice ages, and Milankovich, in the first half of the twentieth century, who refined and completed that theory by including all the relevant astronomical influences. Neither of them had an easy time, although for different reasons. But although neither would live to see it, in the end their efforts would be vindicated by the isotope thermometer.

James Croll was born in 1821, and spent the early part of his life in a little Scottish village. His family worked a small piece of land, but this did not provide enough for them to live from. Croll's father worked chiefly as a stonemason, travelling to wherever there was work, so that young James saw little of him. At the age of 13, James had to leave school and work on the farm, and his formal education ceased at that point. But he read widely, and encouraged by his mother he studied books on philosophy and science. In his book *Climate and Time*, published in 1875 (from which subsequent quotes are taken), Croll later recalled that although self-taught, he reckoned that by the age of 16 he had a 'pretty tolerable' knowledge of the basics of 'pneumatics, hydrostatics, light, heat, electricity and magnetism'. But the farm could not support him, and he had to find a career. Hoping that the work might appeal to someone who enjoyed studying theoretical mechanics, he became a millwright. It was a mistake, for 'the strong natural tendency of my mind towards abstract thinking somehow unsuited me for the practical details of daily work'.

That comment sums up Croll's life. At the age of 21 he gave up the trade of millwright and returned to the family home, to concentrate on studying algebra. To eke out a living while he did so he became a carpenter, and found for once that the work suited him, only to be forced out of it by the lingering effects of an elbow injury he had suffered as a boy, and which was aggravated by the work. Later, he would recall that the injury was what set him on the road to scientific achievement, and that had it not been for the stiff arm 'I should in all likelihood have remained a working joiner'. Instead, he tried a succession of occupations that were easier on his elbow, and gave him more time to read and think. He worked in a teashop, eventually opening one of his own, and took time off from his reading to marry. When the business failed, at least partly because of his obsession with the wrong kind of books, he tried running a hotel. It is an indication of his lack of business acumen that he chose one in a Scottish town which already had 16 inns serving a community of 3500 people, and, an abstainer himself, he refused to serve whisky to his customers.

By 1853 Croll was an insurance salesman, a job he loathed. But four years later that too came to an end when the Crolls were forced to move to Glasgow so that James's wife, who was ill, could be looked after by her sisters. A period of obviously blissful unemployment followed, during which Croll completed a book, *The Philosophy of Theism*, and actually managed to find a London publisher for it. Incredibly, the book was a modest academic success, and it made Croll a small profit. Then, in 1859, he found his true niche in the world.

Croll became a janitor at the Andersonian College and Museum in Glasgow. 'I have never', he said, 'been in a place so congenial to me'. The pay was poor and the work menial, but the job gave Croll two things he prized above all else: time to think, and access to a first-rate scientific library. He began to publish scientific papers, at first on electricity and other problems in physics. By the middle of the 1860s, as geologists became convinced that there had indeed been an ice age, there was a great debate about what might have caused the Earth to cool. Croll became interested, and read up on the astronomical theories, ideas based on changes in the Earth's orbit. He improved on those calculations, and published his first ice-age paper in the respected scientific journal the *Philosophical Magazine* in 1864, at the age of 43. The paper attracted attention, and Croll was urged to take his work further. He did so to such good effect that by 1867 he was offered (and accepted) a post with the Geological Survey of Scotland, and in 1876 he was elected a Fellow of the Royal Society.

The work which enabled a museum janitor to rise to the very top of the scientific tree developed the idea that, as the Earth moves around the Sun, there are long, slow changes in its orientation which alter the amount of heat arriving at different latitudes in different seasons. We see such effects every year.

The Earth is tilted in space, lying at an angle of about 23.5° to the perpendicular to a line joining the centre of the Earth to the sun (this angle itself varies, and is one of the components of the astronomical model of ice ages). When one hemisphere is tilted towards the Sun, it is summer, but at the same time the opposite hemisphere must be tilted away from the Sun, and there it is winter. The difference between summer and winter is very much like the difference between an interglacial and an ice age. Could orbital effects account for that change as well?

Croll calculated how the amount of heat arriving in summer and winter has varied over thousands of years, as the Earth wobbles, like a spinning top, in its orbit around the Sun. He found a regular pattern of changing seasonal temperatures, and he inferred that ice ages would develop in the northern hemisphere when the effects conspired to produce cold northern winters. In fact, this inference is wrong: cold northern winters do not trigger the spread of ice. Paradoxical though it may seem, the improved timescale of ice ages provided by the isotope thermometer, and modern calculations using high-speed computers, show that cool *summers* are more important for the spread of ice. The logic of this is quite straightforward.

With the present geography of the globe, there is always snow in the northern hemisphere in winter, but it melts in summer. If summers were a little cooler, however, some of the snow on the ground might last the whole year round, building up, year by year, into great ice sheets. The more the snow and ice spread, the cooler the world would get, because the shiny surface would reflect away heat from the Sun. An ice age would begin.

After initial interest in Croll's work and the recognition of his achievement in developing the astronomical theory, the idea fell from favour. Partly because Croll had the trigger of ice ages wrong, but also because his model was incomplete, his version of the astronomical theory implied that the present interglacial had begun about 80,000 years ago. But evidence began to mount up showing that the last ice age ended only about 10,000 years ago. Croll's model (which would have predicted the *onset* of severe glaciation 80,000 years ago if he had only thought of the importance of cool summers), slid into obscurity after his death in 1890. Anyone who stumbled across the idea in the early part of the twentieth century would have regarded it as no more than a historical curiosity, not least because the astronomical variations occur on timescales of tens of thousands to a hundred thousand years or so, and by the early twentieth century 'everybody knew' that this did not match the known geological pattern of the ice ages.

Obviously, either the astronomical theory was wrong or the standard chronology of ice ages was inaccurate. Just about the only person who seems to have taken the second possibility seriously was the Serbian astronomer Milutin Milankovich, born in 1879. Unlike Croll, he followed a conventional route through the academic system, emerging with a PhD from the Technische Hochschule in Vienna, in 1904. Even so, there are similarities between the careers of the two pioneers. Milankovich worked as an engineer for five years, but always hankered after more cosmic calculations than the design of a building or a bridge. He was, he wrote in his 1936 book *Durch ferne Welten und Zeiten* ('Through Distant Worlds and Times'), 'on the lookout for a cosmic problem', and in 1909 he took up a post at the University of Belgrade where he taught physics, mechanics and astronomy. It was a step into the backwoods of European science, but like Croll's janitoring job it left him with time to think. Within two years, he found his cosmic problem – he would devote himself to developing a mathematical theory to describe the changing climate not just of the Earth, but of Mars and Venus as well.

The idea was simple, First, he would calculate the amount of heat from the Sun arriving at different latitudes of the Earth in different months today. He would then be able to calculate the climate patterns on distant worlds, without ever having to go to Mars or Venus to measure the temperature. Finally, he could apply the theory to distant times, calculating how the climate of the Earth had varied in the past. But though the idea was simple, its execution was like the labours of Hercules. There were no computers in the second decade of the twentieth century, and Milankovich had to work out all his sums using pencil and paper. The calculations were almost endless, and it took him more than 30 years to complete his life's work. He took it with him everywhere, including on holiday, and worked on it every day. We shall not retrace every step of that labour of love, but shall leap forward and present you with the fully fledged, modern version of the Milankovich model, which so neatly fits the pattern of ice-age/interglacial fluctuations revealed by the isotope thermometer.

All these variations are caused by the changing gravitational tug on the Earth produced by the Sun, the Moon and the other planets, as each heavenly body follows its own path through space. The way the Earth tilts and wobbles as it orbits the Sun, and the way the orbit itself changes slightly from more circular to more elliptical and back again, actually produce three associated rhythmic variations in the amount of heat reaching the northern hemisphere in each season (the total amount of heat reaching the entire hemisphere over a whole year is always the same; all that changes is the way heat is distributed through the year). One cycle is about 100,000 years long, the second is about 40,000 years, and the third is a complicated set of variations around 23,000 years along. Modern computer calculations show how these three varying cycles combine to produce a pattern of variations that closely matches the pattern of roughly 100,000 years of ice age and 10,000 years of interglacial revealed by the isotope thermometer. Mathematical analysis of the fluctuations in temperature revealed by the isotopes show up exactly the same set of three fundamental rhythms, 100,000, 40,000 and 23,000 years long, beating together. If the Milankovich model had not already been worked out when this discovery was made, it would have had to be invented very quickly to account for the discovery. No two ice-age/interglacial cycles are exactly the same, but for 2 million years the real world has marched to the astronomical beat. The Milankovich model has now become fully accepted as the best explanation of the ice-age rhythms.

Although there have been previous ice epochs on Earth, when one or both poles has been covered by land, the ice epoch during which the human line has emerged may be unique. This seems to be the only time a frozen southern continent has been balanced in the north by a nearly landlocked polar ocean. And the Milankovich rhythms can exert their influence strongly because the presence of dry land around the north polar region provides a platform on which snow and ice can build when summers are cool. The Milankovich rhythms may be a unique feature of the current ice epoch, and that explains why another unique feature of our planet today, intelligent life, also emerged during the present ice epoch.

Ice-age people

Whatever the exact cause of the ice-age rhythms, the important point for our story is that during the Pleistocene the Earth has indeed been plunged into a succession of long ice ages, broken by short interglacials. This has never happened before. Instead of the climate changing slowly and steadily over a very long period of time (while those species that are able to adapt to the trend), or changing abruptly into a new pattern (with many species going extinct and the survivors evolving to fill the vacant ecological niches), we have had both effects operating in miniature, repeatedly. The recent pattern has been one of successive recurrence of harsher conditions, broken by short-lived breathing spaces, times of more equable climate. In the heartland of Africa, where our story is still focused, the harsh conditions showed themselves as dry ages, rather than ice ages, with woodlands dying back and both plants and animals in an intensified struggle for survival. The interglacials correspond to wetter intervals, when trees and other plants temporarily flourished and animals found life easier. We believe that this repeated tightening of the evolutionary screw and easing off of the pressure hastened the evolution of the human line by putting cunning and adaptability at a premium, and providing opportunities for an *intelligent* upright ape to succeed in a changing world.

Each time the forests shrank, the apes that were best adapted to the woodland life would continue to do well, deep in the heartland of the surviving forests. Out on the plains, animals that were adapted to the way of life on the savannah would also do well. The creatures that would suffer most would be the ones that inhabited the edge of the forest, including apes that had not yet entirely abandoned their brachiating past. As the forest shrank there would be more competition for space in the trees, and the least successful woodland apes would be pushed further out onto the plains. There, they had to survive or die. Many must have died, with only the most cunning individuals surviving. If the dry age had lasted for a million years, maybe all these reluctant plains dwellers would have been wiped out. But after a hundred thousand years or so, the rains returned and the forests, temporarily, thrived. There was space for the upright apes to recover, both literally, if they retreated to the shelter of the woods, and to gain metaphorical breathing space as food supplies became more plentiful and the ecological pressure on them was reduced. Just as the numbers of their species were beginning to build up again, however, back would come the drought, winnowing out the less adaptable individuals, once again putting a premium on intelligence and adaptability. Repeat that cycle a dozen times or more over a couple of million years, with natural selection taking its toll of the dimwitted each time, and it is little wonder that the survivors of the apes that got pushed out of the woods became quite bright.

In each ice (dry) age, many individuals die. While some species respond by becoming better forest apes, or more efficient plains carnivores, only the most intelligent and adaptable individual plains apes survive, and pass on their successful genes to their descendants. In the interglacials, the descendants spread out. The living is easy, and proto-people, selected for adaptability and cunning, do well. But in the next dry age, once again only the most adaptable – the ones who can cope with harsh, changing conditions – do well. If the ice had come in force 4 million or even 3 Myr, and stayed, East Africa might have turned into a desert, and all the African apes would have died out. Our ancestors survived, and we are here today, because of the unusual pattern of climatic changes they experienced. We are, indeed, children of the ice.

Driven by these climatic rhythms, *Homo erectus* spread out of Africa and around most of the world during the Pleistocene. For a million years the species stayed much the same, apart from a steady growth in brain capacity, indicating the increasing intelligence of the upright ape. But somewhere between about 400,000 and 200,000 years ago there was the beginning of a much bigger increase in brain size, accompanied by a thinning of the skull bones. They had become *Homo sapiens*, modern human beings in all except minor anatomical details. Soon after, the *sapiens* line split. By 100,000 years ago two subspecies existed, although not necessarily side by side. One, bigger built and with the larger brain, lived in western Europe (south of the ice) and across into the Near East and Central Asia. It thrived until about 40,000 years ago, in the midst of the latest ice age, and then disappeared from the fossil scene. It is known as *H. sapiens neanderthalensis*, or Neanderthal Man.

The epithet 'neanderthal' today conjures up for most people an image of a dimwitted, shambling ape-man, but this is an unfortunate misconception and a slur on a very close relative of ours who was both fully upright and intelligent, and thrived in a variety of environments from the edge of the ice cap in Europe to Central Asia. Neanderthals were the first people known to have buried their dead. Sometimes their bodies, carefully laid to rest, were accompanied by valuable flints and stone tools, and food for the journey into the afterlife. At one famous site in the mountains of present-day Iraq a man was buried in a grave filled with spring flowers. Neanderthals were not shambling ape-men, but sensitive and caring people who seem to have disappeared from the evolutionary scene because they were overwhelmed by the even greater success of their closest relations, *H. sapiens sapiens*. The relationship is so close, indeed, that it is possible that the Neanderthals disappeared not by going extinct, but through interbreeding with our own line. It would be nice to think that the genetic line responsible for the floral burial tribute 60,000 years ago in the Zagros Mountains still survives in us today.

Fully modern humans, *H. sapiens sapiens*, were also around by 100,000 years ago, more or less at the beginning of the latest ice age in the present ice epoch. One likely pattern of the emergence of our own variation on the upright-ape theme is that the human form first emerged, as *erectus* had done, in Africa. This is backed up by the discovery of part of a human skull, 115,000 years old, in a cave of South Africa. If this idea is correct, then *H. sapiens sapiens* probably moved out of Africa in its turn, interbreeding with and replacing other varieties of *Homo* derived from *erectus* stock, including Neanderthals. It may be that there were two main waves of *H. sapiens*, first the Neanderthals and then ourselves, hot on their heels. Nobody can be sure exactly how we emerged from the *erectus* line. But although we may never know the details of how *H. sapiens sapiens* finally appeared on the scene, we do know what happened next.

The Neanderthals disappeared about 40,000 years ago in the Near East, and 35,000 years ago in Europe. For the next 25,000 years, while the world was still in the grip of an ice age, it was *sapiens sapiens* alone who represented the *Homo* line. They did well enough even during the last, severe stages of that ice age, as shown by fossil remains, carvings and cave paintings from France and Spain, in particular. Ice-age humans completed the peopling of the world begun a million years before by *Homo erectus*. They spread down through Asia to Australia, and north over the land bridge into America, continents that *erectus* never reached. There is some evidence that a gap in the ice between the Laurentide ice sheet and the coastal glaciers west of the Rockies may have allowed humans to begin to move south into first the northern and then the southern American continents about 25,000 years ago. When the ice melted and the Bering Strait became filled with water once again, they would be cut off from their cousins until the arrival of Viking voyagers from Europe about a thousand years ago. How the Vikings got there is a story that involves, as we tell in Chapter Six, the growth of civilization in the Middle East and then in Europe as the ice age gave way to an interglacial.

The interglacial itself was no different, except in detail, from the dozen or so interglacials that had preceded it since the time of *H. habilis*. It just happened that this was the first interglacial after the winnowing process of ice-age rhythms had resulted in the appearance of a *very* bright ape on the scene. The first humans had already adapted to every continent of the globe, and the worst weather the ice age could throw at them, and were still doing quite nicely, thank you, when the ice began to retreat. So when the more equable conditions of the interglacial arrived, they did not have to use the breathing space to recover from past deprivations. Instead, they exploded into prominence as the number one species on Earth, creators of the first civilizations, with an insatiable curiosity about everything, including their own origins. From now on, the story we have to tell deals with history, not evolution. But the history (and prehistory) of *H. sapiens sapiens* has had a part to play in the evolution and, especially, the extinction of many other species.

CHAPTER SIX

After the Ice: Viking Voyages and a Christmas Carol

If the argument that the rhythmic pattern of ice ages during the Pleistocene helped to speed the evolution of the human line holds water, then you would expect a similar argument to apply to other species. The 'evolution machine' of Pleistocene ice-age rhythms should have been at work on other species as well, producing a diversity of mammals to occupy the diverse ecological niches of the times. That is exactly what we find. Of 119 species of mammal that now live in Europe and Asia, for example, just six were present in the Pliocene. All the rest evolved during the Pleistocene epoch, together with many other species that thrived during the ice age, but are now extinct. Even those 119 species represent only a fraction of the number of mammal species that used to roam Eurasia. In the earliest part of the Pleistocene, new groups that emerged in Europe included true elephants, cattle and ancestors of the zebra. A little later, forms more clearly adapted to Arctic conditions appeared, including woolly mammoths, reindeer, lemmings, musk ox, woolly rhino and moose. The all-time peak of mammalian evolution, in terms of the number of separate genera of mammals alive on Earth, occurred around a million years ago.

As well as encouraging the evolution of a broader variety of mammals than ever before, the Pleistocene ice-age rhythms also encouraged them to spread around the world. When the northern ice advanced, high-latitude species drifted generally southwards, adapting and evolving to slightly different conditions along the way; during interglacials, they spread northwards, and radiated into new niches. But this does not mean that they retraced exactly the steps of their ancestors. The southward migration might actually have been a bit west of south, for one particular group of animals, to one side of a range of mountains; a hundred thousand years later, descendants of those migrants might be moving in a northwesterly direction, up the other side of the mountain range. The climatic rhythm encouraged the spread of mammals east–west, as well as north–south.

Althogh the variety of mammals has declined slightly since their 1 Myr peak, even at the height of the latest ice age, some 18,000 Yr,* the region close to the ice was inhabited by caribou, mammoths and collared lemmings (among other species), while a little farther south the mammal population included mastodons, ground sloths, moose, horse, bison, snowshoe hares, musk ox and a variety of small mammals. But about 15,000 Yr, exactly at the time that the latest ice age was giving way to the present interglacial, a wave of extinction began to hit many of these species, especially the larger mammals. The extinction peaked around 11,000 Yr. Some 39 genera, as many as 70 per cent of the species of large mammal in North America, went extinct at this time, with a smaller, but still large, number of extinctions in Eurasia and Africa. A little later, many smaller mammals and flightless birds disappeared from the Pacific islands and from New Zealand.

Apart from the number of species affected, this wave of extinctions unusually (perhaps uniquely) singled out large animals. Beavers the size of bears, bison with horns that spread a full two metres, and ground sloths *six metres* tall. All were thriving members of the ice-age community, and all disappeared around 11,000 Yr, along with species of elephant and lion that were giants by modern standards, and other gigantic forms of familiar mammals.

It is possible that these extinctions could have been related to the changing climate. After all, they did come just as the Earth was switching from an ice age to an interglacial. But this seems unlikely on two counts: first, such a shift in climate ought to make life easier, not harder, for large mammals; and secondly, there was nothing like this wave of extinctions at the end of any of the preceding twenty or so glaciations. But there was, of

* An abbreviation for 'years ago'.

course, a new factor at work at the beginning of the present interglacial. Human beings had spread around the globe by then. The brainy bipeds must have been efficient hunters, and large mammals would be the obvious first victims of their new-found skills. At the same time, human activities would have changed the environment in which the large mammals lived: cutting off migration routes for these animals, perhaps, or denying them access to water holes that people wanted for themselves. It would, indeed, be surprising if the explosion of human activity at the beginning of the interglacial had not left its mark on other mammal populations. Perhaps the strongest piece of circumstantial evidence that people were the main cause of these extinctions, however, comes from North America, where the extinctions were most pronounced. There, mammals had been evolving throughout the Pleistocene largely free from human interference. When the peopling of the Americas began, species that had no experience of coping with the brainy biped were swept out of the way in the twinkling of an evolutionary eye.

The peopling of the Americas

No two ice ages are exactly the same. The one during which *Homo sapiens* spread around the world almost began 115,000 Yr, with a severe spell of cold weather that lasted for several millennia, but had eased by 108,000 Yr. Instead of a full ice age, there was 'only' an interval of ice-age cold spanning, as writer Nigel Calder graphically expresses it, a rather greater length of time than that which separates us from the builders of the Egyptian pyramids. About 95,000 Yr there was another false start to the ice age, lasting for about 7000 years; but again the climate recovered to something like its present state. Nobody knows exactly why these climatic crises happened (perhaps it had something to do with volcanic activity), but it is interesting that they bracket the emergence of *H. sapiens sapiens* in the fossil record. Between 80,000 and 70,000 Yr temperatures fell once more, partially recovered, and then plunged again. This time, the cold lasted for more than 50,000 years.

Just as there were downward blips of temperature even

during the previous interglacial, so there were millennia of relative warmth even within the latest ice age. The cold bit hard between 70,000 and 60,000 Yr, but slightly less hard from 60,000 to 20,000 Yr. Around 50,000 Yr, there were several millennia of less harsh conditions, and a shorter-lived but even milder run of centuries around 30,000 Yr. These easings of the climatic screw coincide tantalizingly well with the sudden spread of *H. sapiens sapiens* and the disappearance of *neanderthalensis* from the scene, but there is no way of telling for sure whether climate played a major part in these developments, or simply helped the brainy ape who was by now ready to take over the world. Either way, the ice age still had a trick up its sleeve. The greatest advance of northern ice (the 'last glacial maximum') actually occurred as recently as 18,000 Yr, heralding about 4000 years of the worst ice-age weather, before the thaw began. At some time between about 12,000 and 10,000 Yr, the ice age ended and the present interglacial began. Exactly which date you choose to mark the boundary is rather arbitrary. It took thousands of years for the ice to melt, and you might reasonably argue either that the ice age ended when the retreat began or that the interglacial began when the retreat finished. On the basis of the Milankovich cycles, you can even set the boundary date at 15,000 Yr, when the astronomical influences tilted in favour of warm northern summers. For the convenience of having a round number, we prefer to date the beginning of the present interglacial at 10,000 Yr.

Somewhat chauvinistically, geologists also use the end of the latest ice age as the boundary between two epochs, ending the Pleistocene and starting the Holocene (or Recent) where the interglacial begins. There is no physical justification for this. In terms of climate and of the processes operating to shape the Earth, the present interglacial is just one more in the long succession of Pleistocene ice-age rhythms. The only difference is that people are now taking notice of those rhythms, and like to mark such an important event (to us) as the beginning of civilization by saying that it marks the start of a new geological Epoch. But then again, perhaps any geologists around a million years time from now will see traces of the extinctions of large mammals in the fossil record from the time of the beginning of our present interglacial, and it will indeed stand out to them as a geological marker. Human beings may have become a factor in the geological equations as the Holocene began, and (as we shall see in Chapter Eight), human beings may now even be affecting the pattern of ice-age rhythms. Perhaps identifiying the end of the Pleistocene with the emergence of human civilization is not such a bad idea after all.

A combination of climatic changes and human activities certainly seems the best way to account for the extinctions in North America at the end of the ice age. We should perhaps make it clear that by now we are not in any sense talking about 'ape-men' or stupid savages. Cave paintings from 40,000 Yr show a world of sophisticated European hunters and the animals they hunted; a grave site from near present-day Moscow, dated to about 30,000 Yr, shows traces of tailored clothing, including trousers, moccasin-type shoes, jackets and hats, all decorated with beads; the bow and arrow had been invented by 20,000 Yr; millstones from 19,000 Yr hint at the beginning of grain cultivation in the Nile Valley. And while these developments were taking place in Europe and the Near East, the relatives of those early hunters and agriculturalists began to move into North America, across the dry corridor of the present Bering Strait and south from Alaska, through the glaciers. Other people were moving south from Eurasia at about the same time, island hopping their way to Australia, where the oldest known site of human occupation has a firm date of just under 33,000 Yr. Ice-age people were fully human, as intelligent as we are and probably more skilful with their hands, through constant practice. They could talk, they had complex societies, and, judging by the evidence from burials, they had religious feelings and ideas. We have achieved more than them, in terms of bending the world to suit ourselves, primarily because we have the benefit of a basis of hundreds of generations of human endeavour behind us, and because we live in the more benevolent conditions of an interglacial. If each of your immediate ancestors had reproduced by the age of 20 (a reasonable figure for any time before the twentieth century), then just a thousand generations separate you from the people who first populated the Americas. The chances are you've stood in a queue of people that long at some time in your life, heading for a concert or a sporting event; a 5 km tailback of cars would contain a thousand frustrated drivers. And just a thousand people, in a line stretching back in time, not space, link you with those American pioneers.

Of course, people did not just march over the land corridor and head off southward with their packs on their backs. At each stage of the process most people stayed where they were, and made a living from the land near their settlements. Growing populations, and adventurous spirits, would result in new settlements being established, just over the next hill, or down in the next valley. Nobody planned to populate the Americas; it just happened, rather in the way that the first European settlers in the east of the continent, thousands of years later, just naturally drifted westwards in search of peace and quiet, and a bit of land where they could make a living and raise families. The trouble was, when their children grew up they needed some land of their own, so the migration just had to carry on. The first Americans may have been hunters rather than farmers, but the same need applied, especially when overenthusiastic hunting reduced the population of food animals at a particular locality.

Successive waves of immigrants moved into Alaska during the relatively mild conditions from about 40,000 Yrs onward. The experts still argue about details of the timing of these movements, but several Alaskan sites have yielded bone implements that date back to at least 25,000 Yr. How and when the first adventurers risked the journey south through the corridor between the great glaciers can never be determined, but since the gap in the ice closed during the last glacial maximum around 18,000 Yr, it seems certain that only small bands of people can have made the journey before then. Perhaps the descendants of those pioneers lived quietly on the fringes of the ice, following an 'eskimo' way of life, until the thaw began; perhaps no significant numbers of people moved south from Alaska until after the ice began to melt. Either way, when the ice did melt there were people in the northern part of the Americas who were ready to spread southward as the climate changed. The timing of that spread follows the change in climate so closely that it cannot be a coincidence. Quite suddenly, in a short space of time around 11,000 Yr, people spread all across North America, leaving traces in the form of characteristic arrow and spearheads from the Pacific to the Atlantic, and from Alaska to Mexico.

For such recent archaeological remains, the dating becomes very precise. By 10,900 Yr there were people throughout Central America, and beginning to move into South America; it took their descendants only another 300 years to reach and pass the line of the Amazon, and by 10,000 Yr even the tip of the continent was occupied. In a thousand years, exactly at the end of the ice age, humans spread from Alaska to Cape Horn, wiping out many mammalian species along the way. Half a world away, the same climatic changes that led to the peopling of the Americas were encouraging the development of a new way of life that became known as civilization.

The warmth and the wet

Hardly surprisingly, the warmest part of the present interglacial was just after the end of the latest ice age. With the present-day geography of the globe, it takes all the Milankovich rhythms working together to pull the world out of an ice age and start an interglacial. Once the ice has melted, the rhythms may begin to get out of step with one another, weakening the warming influence. But even if the planet cools off a little as a result, the ice age will not return immediately because the bare ground and open sea revealed by the retreat of the ice is more efficient at absorbing heat from the Sun than the reflective ice sheets were. So the typical pattern of an interglacial is to start with a burst of warmth, and then to slide slowly but inevitably, with minor ups and downs of temperature, back towards conditions which will allow the ice to advance once again. Each mild spell during the interglacial is less mild than the one before; each cold snap is harder than the last burst of cold, until one day a cold snap starts and does not end for a hundred thousand years.

The early Holocene was warm and wet. It rained a lot because the increase in temperature meant that there was more evaporation from the oceans (and inland lakes and seas), increasing the amount of water vapour in the air and making more clouds. Increased rainfall encouraged the spread of northern forests over what had been dry grasslands. It was also wet in regions where the ice was melting, a process that did not happen overnight or in a single year, but which took centuries, even millennia, to complete. And as ice melted and the sea level rose, it was pretty wet in coastal regions around the globe, with land that had been dry for at least 70,000 years flooding as the sea returned. The flooding was even worse in the north. Huge areas of Siberia, for example, had literally been pressed down by the weight of ice above, sinking into the treacly layers of molten rock below the surface. When the ice melted the weight was relieved, but the land could only lift back up to its former height very slowly through the clinging, treacly magma; indeed, the process is still going on now, 10,000 years later. The Arctic Ocean flooded in over the depressed land surface.

At the same time, other areas of water, great inland lakes and freshwater and salty inland seas, were drying up. During the ice age, partly because of shifts in the wind patterns and rainfall belts, and partly because water from the edge of the ice sheets could evaporate in summer to make clouds and bring rain to regions that are now far from moisture-bearing air streams, there were enormous numbers of lakes south of the ice in regions such as North America and Europe. The Great Salt Lake of Utah is a remnant of an ice-age sea, known as Lake Bonneville, that covered an area of 50,000 km2 to a depth of 300 m, and there were other inland seas about half as big in northwestern Nevada and southeastern California. The Caspian Sea was twice as big as it is today, spreading into central and eastern parts of modern Russia, and still comparably big even when the ice began to retreat and northern Sibera was flooded. The lakes themselves may even have grown at first, as the ice melted back and there was more moisture in the air.

The effects of all this on human populations were dramatic. The remains they have left behind show that early populations of people preferred to live near the seas. Fish may have been an important part of their diet, and they had, it seems, learned the trick of preserving food with the aid of salt, which they could get from sea water. Ice-age people were also seaside people, certainly around the Mediterranean and the North Sea, and along the Great Australian Bight, as well as in other parts of the world. Some researchers suggest that the main centres of these early human populations were in regions that are now submerged beneath the waves; it is argued that the immediate efect of the climatic changes and coastal flooding at the end of the ice age may have been to *reduce* the number of human beings in the world, and that the accompanying disasters are the origin of the legends of a great flood that have come down to us from ancient times. From 15,000 to 10,000 Yr, the global sea level rose by at least 50 m; by 5000 Yr it had risen a further 40 m. Since then it has never been more than a few metres above or below its present level.

The warmth and wetness of the early Holocene – a time sometimes referred to as the 'postglacial optimum' – were at a peak between about 7000 and 5000 Yr. The world was generally about 2°C warmer than in the middle of the twentieth century (perhaps 10°C higher than temperatures had been during the last glacial maximum), and there is virtually no evidence of any major deserts at that time. Rivers flowed in the centre of what is now the Sahara, and other modern desert regions, such as the Thar, or Great Indian Desert, had twice or three times as much annual rainfall as they have had in recent years. Such changes encouraged the development of human activities around the world once the shock of the initial postglacial flooding had been overcome.

Into and out of the Fertile Crescent

Archaeological evidence from the caves of the Zagros Mountains, which lie across the borders of modern Iran, Iraq, Turkey and Syria, show that people had begun to herd sheep and goats and to cultivate grain by 11,000 Yr. Not far away, and at about the same time, the world's first city (we would probably refer to it as a village today) was being built at the site of Jericho, in the Jordan Valley near the Dead Sea. The Dead Sea was one of the great ice-age inland seas that was now in retreat, drying up and leaving behind great deposits of salt. This salt, the reason for the city's existence, was a valuable resource used not just for preserving food but also in tanning leather and in baking.

Agriculture seems to have developed independently in three main centres around the world. The first, and the one we shall concentrate on most, was in the so-called Fertile Crescent, sweeping from the eastern Mediterranean region around Jericho and the Zagros Mountains, east to the valleys of the Tigris and Euphrates rivers and down to the head of the Persian Gulf. The second centre of agriculture was in China, where rice, millet and yams (among other plants) were cultivated, and pigs herded, from about 7000 Yr. The third centre was in Central America, where people were cultivating crops such as maize by 5000 Yr. The warmth and the wetness encouraged humans to begin their experiments with the farming way of life. But, almost paradoxically, what may have concentrated their attention on farming *as* a way of life was the retreat of the climate from these so-called optimum conditions.

After about 5000 Yr the world cooled slightly and, as usually happens when the world cools, the climate of mid-latitudes became drier. Glaciers advanced in the Alps for the first time since the ice age had ended, and the Sahara and other great deserts of the present day began to make their appearance. Hubert Lamb, a pioneering British climatologist who has championed the idea of climatic change as an influence on human affairs in historical times, has gathered together evidence for the way these changes affected the people of the time following the postglacial optimum. By 5000 Yr elephants and giraffe were becoming rare in Egypt, and by 4500 Yr they had disappeared altogether from the region, along with rhinos. The annual flooding of the Nile, fed by seasonal rainfall over Ethiopia, reached lower levels than it had in the previous millennium, but at the same time this reliable source of water became increasingly important to people as the rest of the region dried out. Exactly the same processes were at work in the Fertile Crescent, which in the third millennium BC was no longer fully living up to its name. Decreasing rainfall and the spread of desert regions outside the river valleys made the Tigris and Euphrates increasingly important as sources of water and of life. Hunting became less and less reliable as a means of obtaining food, and even gathering wild crops could no longer be relied upon. The more the climate deteriorated, the more people had to turn to the relatively new invention, agriculture, to support themselves.

The same process, suggests Lamb, was occurring in India and

China. He mentions too the proposal by a Japanese meteorologist, Hideo Suzuki, that refugees from the increasingly arid regions nearby may have become the slaves that made intensive agriculture possible, and whose labour enabled the Egyptian pyramids to be built. Biblical accounts of the wanderings of the Israelites in the wilderness, in a search for a new home, stem from exactly this period of upheaval and mass migrations in the Near and Middle East. People were squeezed into regions such as the Fertile Crescent by the change in climate, and there they learned to become more efficient farmers than ever before. The seeds of modern civilization had been sown, not by the ice age itself but by a shorter and less severe spell of cold, a mini ice age. The children of the ice had learned their lessons well, and coped more than adequately. As the new farmers learned to cope with the conditions, and civilizations became so well organized that surplus food produced on the farms of fertile land could be transported and used to support people in other places who did not farm for a living (city dwellers and armies, for example), the squeeze into the Fertile Crescent rebounded into an outward surge of civilization.

Western Europe became the dominant force in world history as a result of this surge. China became a great civilization first, and remained civilized for the longest continuous span of time. But by and large China, for cultural reasons, kept itself to itself, or at least to its own corner of the globe. Central and South America developed great civilizations later, but these had no time to spread beyond their home continent before they were overrun by other cultures. It was the Europeans that did the overrunning, having developed the culture that eventually took itself to every corner of the world. And that expansionist tendency was already apparent in the early days.

The colder climate brought with it a change in weather patterns that provided reliable winter rainfall but warm summers in the eastern Mediterranean and along the northern fringe of Africa, even while the deserts were spreading further to the south and east. In the north, although there were no spreading hot deserts, conditions were just as bad as in the arid zone. Glaciers advanced until, shortly after 1100 BC, they were as far south as they have ever been during the present interglacial. In North America there were no glaciers south of the present Canadian border during the postglacial optimum; all the glaciers on the US Rockies today have formed since 1500 BC. But while no stories of those cold days have come down to us from native American legends, Scandinavian storytellers incorporated the story of those icy days of 3000 Yr in the legend of Ragnarok, the twilight of the gods, when three severe winters followed each other in succession, without a summer in between: the Fimbulvinter that heralded the end of the world. The great conflagration that ends the story may, Lamb has suggested, have been as real as the advance of the ice, a folk memory of huge and horrifying forest fires that engulfed the dried-out husks of trees that had been killed by the cold (elements of this story, which we now believe was founded on fact, were borrowed by Wagner for his Ring Cycle).

With all that going on in the north, and the deserts spreading to the south, the place for civilization to develop as it moved out of the Fertile Crescent just had to be the eastern Mediterranean. Egypt, Greece and Rome all in their turn held centre stage; but it was Rome that first took civilization, if not to every corner of the world then at least to every corner of Europe.

The Romans were lucky. Legend tells us that the city of Rome was founded in 753 BC. For five hundred years Rome was a minor city state in a Mediterranean world dominated by the activities of Greeks, Phoenicians and Carthaginians. It was still relatively cold then, with frosts and snow in Rome itself, at least in some years. Beech trees grew near Rome in 300 BC. But as Rome began to grow in importance, the climate improved slightly. The beech trees, which prefer cooler conditions, retreated to higher latitudes, and frost and snow in the city itself became a thing of the past. The great spread of Roman culture 'coincided' with an alleviation of climate, a retreat from the severe cold that had been at its worst about 2500 years ago, around 500 BC.

The slight improvement and warming in Europe continued until about AD 400, giving the Roman Empire a climate that was distinctly less harsh than the European climate of the middle of the twentieth century. As the Empire spread northward the Romans took grape cultivation with them, introducing vines to England and Germany, to such good effect that by AD 300 the province of Britain had become self-sufficient in wine (probably the most tangible image, to anyone who has experienced recent English weather, of the better climate enjoyed by the Romans). But neither the weather nor the Roman Empire were to last. The Empire fell and the climate declined again (almost certainly a coincidence of timing; there is no sign of any climatic changes in the first millennium AD drastic enough to harm the Empire significantly, even if wine production in Britain did go into decline). There were to be just two more major changes in climate before the twentieth century began. It was to be during the first of these climatic extremes, a period of warmth known as the little optimum (since it was warmer than today, but cooler and much shorter than the postglacial optimum) that Europeans first began to voyage far afield. Although those early voyages did not result in the establishment of a permanent foothold for European culture on any other continent, they did result in one landmark event, the details of which are, alas, lost to us forever. Some time towards the end of the first millennium AD European adventurers from Scandinavia set foot on the mainland of North America and met face to face with other members of *H. sapiens sapiens*, the descendants of an earlier wave of adventurers who had travelled overland into America through Alaska during the ice age. *Homo sapiens sapiens* travelling west around the globe had made contact with *Homo sapiens sapiens* travelling east around the globe, and there were no new lands left that had not felt the tread of the brainy biped. As it happens, vines come into that story, too.

Medieval warmth

Almost as if to prove that the onset of warm weather does not always cause an immediate upsurge in the development of human civilization, the collapse of the Roman Empire in the west (the eastern Empire, Byzantium, was to survive for many more centuries) was followed by an interval of several centuries in which the world became warmer, but European civilization was in such a sorry state that the period is sometimes referred to as the Dark Ages. In some regions of the globe, in some decades, the warmth may have been almost as great as that of the postglacial optimum; but this little optimum, or medieval warm period, was more patchy, both in space and time. Different regions of the globe were at their warmest at different times, and the warmth seems to have passed some places by altogether. In some places, including parts of the Arctic and equatorial latitudes, the warm interval was more or less unbroken from about AD 400 to 1200. But in Europe and North America, the regions of special relevance to our story, the interval from about 650 to 850 was quite cool, with some very severe winters, while in China and Japan these decades were the *only* period of sustained warmth between 400 and 1200.

In Europe and around the North Atlantic, the best years of the little optimum were from about 1000 to 1200. There are many historical records that hint at the differences in climate in those days compared with recent decades. In the year 873, for example, plagues of locusts, thriving in the dry heat, reached from Spain to Germany; in the autumn of 1195 they penetrated into what is now Hungary and Austria. But just as Europe had been relatively cool during the decades when China and Japan enjoyed their modest version of the little optimum, when the optimum was at its height in Europe, China and Japan suffered severe cold. Detailed records survive from these times, and they were analysed carefully by the Chinese climatologist Chu Ko-chen in the early 1970s. During the eleventh and twelfth centuries, the climate of China deteriorated so much that the records describe snow falling a month later in spring than the latest snows of the twentieth century. Plum trees that had thrived in the northern part of China during the warmth of the previous few centuries died out there, while frost killed lychees in the south of the country. Japanese records of the date on which the flowering cherry bloomed each year show that by the twelfth century spring was two weeks later, on average, than it had been in the middle of the ninth century.

Lamb explains these coincidental – but opposite – shifts in climate on opposite sides of the globe in terms of the mass of cold air over the Arctic bodily moving sideways, off the pole. More or less steady winds blow in a roughly circular, zigzag path from west to east around the polar region (which is why it takes longer to fly from London to New York, against the wind, than from New York to London, with a tail wind). These winds (which at high altitudes, where they are steadiest, are called the jet stream) partially isolate the mass of cold air over the Arctic itself. The circular winds, blowing around the globe, are also known as the circumpolar vortex. Because of the influence of mountain barriers in the path of these winds, and in response to other changes in the circulation system of the world (perhaps linked with hanging patterns of sea surface temperatures), the whole vortex and cap of cold polar air can shift across the Arctic region, and may be centred, in any one year or decade, on either the Pacific side of the pole or the Atlantic side. From 1000 to 1200, say Lamb, the vortex must have lain on the Pacific side of the pole. That would have produced a northward shift in the climatic zones around the Atlantic, and a southward shift in the climatic zones around the Pacific, bringing cold to China, Japan and the whole northern Pacific region, but leaving Europe, North America and the North Atlantic basking in the warmth of the little optimum. Nobody knows why the vortex should have shifted in this way at that time, or even if there was any particular reason for the shift. The vortex has, after all, got to be centred somewhere in the Arctic region, and maybe it just happened to be over by the Pacific in those centuries. The reasons for this shift in climate are not our concern here; what does concern us is its effect on the human population of Europe, and especially on the Viking voyagers of the Norse lands.

During this peak of the European little optimum, vine cultivation extended up to 5° further north in latitude and 200m further above sea level than at present, suggesting that average temperatures were a little more than 1° higher in those days. But climatologists and historians are no longer dependent on such proxy records of climatic change once we come to within a thousand years of the present day. The ice sheets over Greenland, laid down as layers of snow falling each year and accumulating on top of each other, contain frozen in their heart a direct record of the average temperature of the North Atlantic region each year. By drilling into the ice and extracting a long core of these layered ice sediments, which can be dated by counting the layers downward from the surface (rather like dating a piece of wood by counting its rings), climatologists can read this record, a frozen thermometer that reveals how temperatures have changed, if not quite from year to year then certainly from decade to decade.

Apart from the not inconsiderable difficulties in drilling a core of ice 404m long, extracting it from the ice sheet and counting the layers of ice to provide a calendar going back 1420 years, the Danish researchers who first obtained such a record of variations in North Atlantic climate since the sixth century AD had to measure the proportions of different kinds of oxygen atoms in the ice from different layers in order to take the temperature of the world when that ice was falling as snow. This is not easy. The trick is essentially the same as the technique used to calibrate the Milankovich rhythms, and depends on the fact that there are two forms of oxygen – isotopes – in the air and in the water (H_2O) of our planet: oxygen-18 and oxygen-16. Oxygen-18 is heavier than oxygen-16, so some water molecules are heavier than others. Lighter molecules of water evaporate more easily from the ocean to form clouds and, in the far north, those clouds produce snow. The exact proportion of oxygen-18 that gets into the clouds (and therefore into the snow) depends on the average temperature over the ocean surface in a particular year. So, measuring the ratio of oxygen-18 to oxygen-16 in the ice samples reveals directly the temperature in the year that ice formed from freshly fallen snow.

Temperatures measured in this way can be checked against historical records for the past few decades, corresponding to the top layers of ice, and this shows that the frozen thermometer is indeed a good guide to temperatures of the past. The record shows that in and around Greenland the little optimum was in full swing by the early part of the eighth century AD, that temperatures fell for a couple of decades in the middle of the ninth century, and that the warmth then returned and persisted, with only minor fluctuations, until well into the eleventh century. Some more erratic fluctuations in temperature then heralded the onset of a long spell of cold centuries, lasting right up until about a hundred years ago (more of that in Chapter Seven). But such dry statistics have no way of conveying what the changes in climate meant to the people who lived in the region at the time. One of the most fascinating features of the temperature pattern revealed by the measurements of oxygen ratios in the Greenland ice is the way it meshes with, and provides new insights into, the limited historical accounts of the spread of Norse culture around the northern rim of the Altantic during, we now know, the little optimum itself. Cultures in the heartland of western Europe may to some extent have missed the chance to develop during the optimum, but the Viking voyagers made no such mistake, and took full advantage of the opportunities provided.

Why Greenland is not green

Historians sometimes try to distinguish between the terms 'Viking' and 'Norse'. The Vikings were seafaring adventurers, tough and hardy. Some were pirates and robbers who looted, raped and pillaged around the coasts of Europe. They were the more disreputable representatives of the Norse culture, which back home in Scandinavia ran to respectability in the form of 'proper' nations, kings and the rule of law (if a somewhat harsher form of law than we are used to today). Vikings were often lawbreakers who had been exiled from their homelands. But the distinction quickly becomes blurred. Some of the Viking plunderers settled in the lands they had been raiding, including England, Ireland and a part of France that is known to this day as Normandy (the land of the Norsemen). Some of them became kings and nobles in their own right: rulers, in a more or less lawful fashion, of their own lands. Some even experimented with democracy. The government which some claim to be the oldest continuous democracy (of a sort) in the world is to be found today on the Isle of Man, and is directly descended from the Norse form of parliament (only 'sort of' democracy because, among other things, women are still excluded from the vote). It was Vikings-cum-Normans who invaded and conquered England in 1066 (and their success was in no small measure down to the fact that the English army was exhausted from its efforts at beating off an invasion of northern Norsemen). The people who spearheaded the spread of Norse culture westwards across the northern rim of the Atlantic were certainly Vikings. But in a generation or two their descendants became respectable Norse men and women, offering allegiance to the homeland and to the Christian religion, not just to the old northern gods. But the colonies of the Westvikings flourished only as long as the climate permitted.

The Vikings were not the first people to venture into the broad Atlantic. Irish monks were the pioneer Atlantic voyagers, seeking peace and solitude away from the crumbling civilization of the Dark Ages and the hordes of barbarians overrunning Christian lands. Their stories have become distorted and undoubtedly exaggerated in the retelling down the centuries, but there must be at least a foundation of truth in the story of St Brendan and his travels in the sixth century. Perhaps he reached America, perhaps not; he certainly got close enough to Greenland to encounter icebergs. When the Vikings established a base on Iceland in the ninth century, they found the Irish monks had got there before them, but the monks quickly departed, now that even this lonely outpost of Europe had been overrun by just the kind of people they wanted to avoid.

The story of the Viking voyages has also been embroidered down the centuries, but large chunks of what seem to be genuine history are included in the sagas, the oral histories that were written down and preserved centuries later. The *Landnam Saga* tells of the first settlement in Iceland, and the *Greenlander Saga* tells of voyages and settlements farther west. One of the pieces of evidence that shows how much truth there is in the sagas is the way in which the climatic events they describe match up with the record of the icy thermometer from the Greenland glacier.

Some time in the middle of the ninth century, on two separate occasions Norse voyagers were blown off course and 'discovered' Iceland, complete with its colony of Irish monks. The monks had eked out a living in a very harsh environment. An early record written by the monk Dicuil in Ireland in 825 records a visit to Iceland, and says that only a day's journey further north from the home of the Icelandic monks there was a permanently frozen sea. The first attempt at a Norse settlement, led by a farmer, Floki Vilgerdason, in the 860s, found conditions even worse. As we now know from the thermometer in the ice, Floki could scarcely have picked a worse time for his adventure, just at the end of the cold decades that split the little optimum in two in the high North Atlantic. He lost his cattle in a severe winter and, as the *Landnam Saga* records, came home to Scandinavia with tales of 'a fjord filled up by sea ice'. And so, the saga continues, 'he called the country Iceland'. Ironically, that is just about the last mention of sea ice in this connection for three hundred years. In the 870s the North Atlantic was warming up, and other settlers following in Floki's wake found Iceland much more hospitable, and established a thriving colony.

Over the next couple of centuries, in the warmth of the North Atlantic optimum, Norse travellers voyaged to the Mediterranean, trading with Italy and with Arab countries; others moved far into modern Russia, always following the great river systems (the Norse were, indeed, involved in founding the state that was to become Russia); and some followed the rivers south and east to Byzantium. From 900 to 1100 Europe belonged, if it belonged to anyone at all, to the Norse. And they very nearly established permanent colonies in America, as well.

Floki and the farmers who followed him had not been Vikings in the true sense of the word, although they must have been rather tougher than the average European farmer today. But the next stage in the sagas of the Westvikings fully lives up to the bloodthirsty image evoked by the Viking name. In 960, back in Norway, a rather nasty piece of work called Thorvald Asvaldsson killed a man and was forced to flee to Iceland, taking his family with him. By now, a hundred years after Floki's ill-fated voyage, the settlement was well established, and the good land in the south of the island was all occupied. Thorvald had to make do with poor land in the north. But his son Erik married into a good family and set himself up on a better farm. He seemed set for a secure life in Iceland when a violent streak to match that of his father surfaced. Outdoing Thorvald, Erik killed two men, and in 982 he was banished from Iceland for three years to give him time to cool off. The sagas refer to him as 'Erik the Red', and it is tempting to see this as an indication of his violent temper, but it may just be because he had red hair.

With a shipload of followers, Erik headed west, deciding to use his period of exile to explore a region that he had heard vague stories about, islands to the west of Iceland that had been seen only by lost voyagers more anxious to return home than to explore what they had found. The land he found was, by and large, rough and rugged. But there was a deep fjord on the southwestern coast, well protected from the sea, warmed by the Gulf Stream and with adequate land for farming near the coast. Conditions were rather like those he had left behind in Iceland, and Erik called the new land 'Greenland'.

Just why he chose that name, we may never know. According to one version of the story it was because from out to sea Vikings could see sunlight glinting green on the glaciers of the Greenland mountains. A more plausible version, recorded in some of the sagas and certainly fitting what we know of Erik's character, is that he chose the name as a deliberate confidence trick, part of a great real-estate swindle to persuade Icelanders to join him in founding a new colony where Erik would be in charge and there would be no threat of further exile. But those written versions of the sagas were recorded centuries later, when the climate of Greenland was even harsher than we know it today, and it must have seemed to the chroniclers that nothing but a swindle could have persuaded anyone to follow Erik west. We now know that Erik arrived in Greenland near the end of a particularly warm part of the little optimum, and that the coastal region where he landed must indeed have been green and fertile, by the standards Icelanders were used to. To be sure, there may have been an element of exaggeration in Erik's sales talk to the people back home, but it was only exaggeration, not an out-and-out lie.

If Greenland and Iceland had been discovered in the same year by the same explorer, then he might quite logically have called Greenland 'Iceland' and Iceland 'Greenland'. One of the islands is indeed covered by ice, and one is more green and fertile than the other. But they got the wrong names, largely because of minor climatic fluctuations: Iceland was settled at the end of a cold spell, and Greenland near the end of a warm spell. That is why Greenland is no longer really green in the agricultural sense.

Once they had settled Greenland, it was probably inevitable that the Norse would reach the mainland of North America. Land west of Greenland was discovered as early as 986, by Bjarni Herjolfsson, a merchant whose ship was blown off course on one of the first trips from Iceland to the daughter colony. But Bjarni made no attempt to land in the new country.

He was a merchant, not a Viking, short on supplies and with a cargo to deliver to its destination. Indeed, there was so much work to be done building up the settlement in Greenland that it was not until the middle of the 990s that Leif Eriksson (the son of Erik the Red) set off to explore the new territories. Just one settlement, at L'Anse anx Meadows in northern Newfoundland, has been discovered and investigated by archaeologists; but that is enough to confirm that Vikings did reach North America at the end of the tenth century. Another settlement farther south is mentioned in the sagas, which recount tales of Vinland, a wonderful, warm and fertile place where fruit grew on trees for the taking, game was plentiful and there was no need to work for a living. The stories, undoubtedly, were exaggerated; but like the naming of Greenland, they must have had some foundation in truth.

The main resource that the Vikings developed in Vinland, wherever it was, was timber; wood was in desperately short supply in Greenland and Leif became both rich and famous from his travels in the west, earning the name Leif the Lucky. But while the climate was benign, there was another hazard for settlers on the American mainland to face, which possibly explains why any colonies that were established to the south and west failed to survive. North America, unlike Iceland and Greenland, was already inhabited. Humanity had closed the circle around the world when Viking voyager and native American met face to face. By 1066, when William the Conqueror invaded England from Normandy, Norse and Amerindians had already met.

It is tantalizing, but futile, to speculate on how that contact might have developed if communications across the chain of colonies back to Europe had remained unbroken. Would the Europeans have engaged the inhabitants of America in bloody war? Or, more likely in view of the long and difficult chain of communications, would they have established friendly trading relations, leading to the development of an American culture that would be well able to withstand the impact of increasing contact with Europe as ship-building technology improved and direct voyages across the Atlantic became possible? We shall never know. What we do know is that this contact marked the limit of Westviking expansion. Within a few decades the climate began to deteriorate; the colony in Greenland went into decline and eventually died out. Any Norse left on the mainland of America had to fend for themselves; their settlements also died out (or perhaps, we would like to think, were absorbed into native American tribes). Even the settlements in Iceland survived only by the skin of their teeth through the worst ravages of what became known as the little ice age. And on the mainland North America a civilized *native* way of life was destroyed. From now on, the Norse voyagers and their colonies have little part ot play in our story. Their tale is very much a might-have-been, and after the end of their story the focus shifts to Western Europe and North America. But we owe it to those pioneers to discuss their fate in just a little more detail, not least because there are important lessons to be learned, even today, from the failure of the Greenland colony.

The return of the ice

When the ice came back to Greenland in full force, the Norse colonies were doomed, not because life in that part of the world became impossible but because they failed to adapt their lifestyle to the changing conditions. The Greenland colonies survived for about five hundred years, roughly from 1000 to 1500, so they were far from being a complete failure. Erik's original colony at the southern tip of Greenland was known as the Eastern Settlement; another colony, further north on the western side of Greenland was known as the Western Settlement. Over a large part of that 500-year span the Norse in the Western Settlement, in particular, were in contact with the Arctic hunters, generally referred to today as 'Eskimo', but in this particular case members of the Inuit people. The lifestyles of the two cultures could hardly have been more different. The Norse were settled people, with farms and cattle, who also used the nearby sea as a resource. The Inuit were nomadic wanderers who followed their food across the ice and would live anywhere they could find that food, just as they had done since their ancestors moved into America all those thousands of years before, during the last ice age.

The Norse colonies were small – perhaps 5000 in the Eastern Settlement and 1500 in the Western Settlement – but they did well as long as the weather remained benevolent. Migrating harp seals, which arrived on the west coast of Greenland each spring, were an important source of food, but the summers were warm enough and long enough for grass not only to provide immediate pasture for the cattle, but to yield a surplus of hay on which the animals could be fed during the long winters. Polar-bear skins and walrus tusks were traded eastward, to Iceland and Scandinavia, in exchange for items the colony could not produce for itself. The most dramatic example of this, and of the temporary wealth of the colony, came in 1125, when the Greenlanders 'traded' a live polar bear for a Norwegian bishop, who they ensconced in grand style in a stone-built cathedral with its own home farm (by then, the colony was more than half as old as the present USA). Two hundred years later, the church owned about two-thirds of the best grazing land on the island (and the colony had then been going for a hundred years *longer* than the USA today). But this very European structure to the Greenlander's society was probably their undoing.

When the North Atlantic cooled in the thirteenth and fourteenth centuries, the Greenland colonies were affected in many adverse ways. As the sea ice spread southwards, it became more difficult and dangerous for boats to make the voyage from Iceland. At the same time, with the expansion of their hunting zones, the Inuit moved southwards, coming into more direct contact – and conflict – with the Norse. And on the farms of Greenland, summer was now too short and wet to provide enough hay to see all the cattle through the winter. Even the seals seemed to have changed their migratory habits as the climate changed, removing another essential resource from the colonists.

In the face of all this, the Norse carried on their traditional way of life as best they could and for as long as they could. There is almost a sense of epic tragedy in the decline of the colonies. The Greenlander's last bishop died in 1378, and was never replaced; there was no official contact with the colonies at all after 1408, although occasional ships would put in to trade or seek shelter from the weather. Archaeological studies show how the surviving members of the shrinking community carried on farming and raising cattle, but the eloquent testimony of skeletons from the graveyard shows that, as conditions became harsher and food more scarce, the average height of the Greenlanders declined from about 177 cm in Erik's day to about 164 cm by the 1400s. Intermittent contact was maintained with the colonies during the fifteenth century. The last bodies laid to rest in the graveyard, preserved by the frosts of even more severe weather that followed, were dressed in styles from Europe from about 1500. But early in the sixteenth century, the last colonist died. In 1540, ships driven to Greenland by severe weather found nobody left alive, and one dead man lay frozen where he had fallen. The last of the Norse Greenland colonists may have missed rescue by only a short time. To put all this in perspective, that last Norseman was closer to us in time than he was to Erik the Red.

But the tragedy need never have happened. After all, the very conditions that killed off the Norse colonists allowed the Inuit to thrive. Thomas McGovern, of Columbia University in New York, has shown how the Norse could have survived, if they had abandoned cattle-keeping and concentrated their activities on developing marine resources; in effect, by adopting and adapting the Inuit lifestyle. Farmers who struggled to the bitter end to keep their herds of cattle alive could have done better simply by changing over to goats and sheep, creatures better able to fend for themselves, and using the extra time this gave them for fishing. An even more radical possibility, says McGovern, is that even in the twelfth or thirteenth centuries the Greenlanders could have developed the kind of whaling, fishing and sealing villages that are characteristic of modern Greenland.

One reason why the Norse followed neither of these options is that they had no wood to build boats. Greenlander boats were traditional, wooden ships like the ones their ancestors had sailed from Norway to Iceland and beyond. As these old boats wore out, the colony became isolated. And yet the Inuit got around in boats built by a completely different technique, out of skin. Equipped with skin boats, says McGovern, the Norse in Greenland could have spread themselves out in little villages and homesteads along the coast, trading with each other and maintaining communications by sea. The Greenlanders did not even have enough sense to adopt the warm clothing the Inuit wore to protect themselves from the cold. The stories of frozen bodies dressed in the European style of 1500, and the last lonely Greenlander, may bring a tear to the eye, but it ought to be one of exasperation. European fashions of 1500 were not suitable clothes to wear in the Greenland winter of the sixteenth century. 'Single-minded conservatism', says McGovern, may have been the single most important factor in the Norse extinctions in Greenland (see his contribution to *Climate and History*, edited by T. M. L. Wigley, M. J. Ingram and G. Farmer). You can almost see that last Norseman, muttering to himself, 'if cattle farming was good enough for Erik, then it's good enough for me'.

This is a most important lesson, since the climate did not stop changing in the sixteenth century. Climatic fluctuations are still occurring today, including some changes in the climatic pattern that result from human activities. In many parts of the world, people are even now being faced, for different reasons, with the same choice that faced the Norse in Greenland seven hundred years ago: adapt, or die. The choice is real. Even a 'primitive' culture can adapt to the rigours of a little ice age: witness the Mill Creek 'Indians' of North America, contemporaries of the Greenland colonists but who did not make the mistake of single-minded conservatism (we look at their example later in this chapter). Will we be less flexible, and less successful, than those long-gone Amerindians?

Although the Norse were the people who seem to have taken the fullest advantage of the opportunities provided by the little optimum, it would be wrong of us to give the impression that the benefits passed the rest of Europe by entirely. The warmth in Europe seems to have continued until about 1300, a little later than in Greenland, and to have coincided with the awakening of the form of European civilization that has continued to the present day. From the middle of the eleventh century onwards there were increases in population and improvements in agriculture, producing wealth which led to a great phase of cathedral-building and to the Crusades, Europe's attempt to wrest control of the Holy Land from

Muslim hands. William's conquest of England, and the subsequent development of a new kind of English culture, were a minor part of all this activity. But by 1300, or even a little before, this phase of European expansion was at an end.

Lamb has suggested that the beginning of the end of the little optimum, may have been responsible for the emergence of the Mongols out of Asia early in the thirteenth century. Just before Genghis Khan and his hordes swept out of Asia, the usually arid heartlands of the continent had, for a time, been relatively moist and productive, as part of the climatic pattern associated with the optimum. Populations increased as a result. The rapid outward expansion of that population, into European Russia on the one hand and as far as Beijing on the other, took place just when high latitudes began to cool rapidly, and there was a great southward advance of sea ice near Iceland. There could, says Lamb, have been an invasion of cold Arctic air into the heart of Asia as part of this new pattern of climate, reducing the productivity of the land and forcing the people to follow their charismatic leader in a quest for survival.

This, as Lamb acknowledges, is speculation: there is no direct proof that climatic change was an underlying cause of the outburst of the Mongol hordes. But their homeland was close to the region of China where there had been severe cold for some centuries (while the North Atlantic enjoyed its little optimum), and there is good evidence that, at the end of the little optimum, cold spread gradually westwards from China to Europe. Lamb's Genghis Khan scenario exactly fits that picture. That wave of cold does not seem to have been caused simply by a shift of the circumpolar vortex to a new position, but by an expansion of the whole vortex as well. This eventually brought cold to high latitudes everywhere in the northern hemisphere: a little ice age, that lasted from the middle of the fifteenth century to the middle of the nineteenth century, and was at its worst in Britain and Europe in the seventeenth century. But as the climate deteriorated from the peak of the little optimum and slid into the trough of the little ice age, it was damp and disease, not biting cold, that first set back the growth of western European civilization, in the fourteenth century.

Death and desertion

The setback in the development of European civilization was heralded by an increase in stormy weather. On four separate occasions in the thirteenth century, disastrous sea floods took the lives of at least 100,000 people in Holland and Germany; the worst of these floods killed more than 300,000 people. Between 1240 and 1362, more than half the agricultural land (60 parishes) in the then-Danish diocese of Schleswig were swallowed by the sea; in other coastal regions sand, not water, was the problem, as the strong winds created marching dunes that enveloped many coastal villages and townships, including the port of Harlech on the west coast of Wales.

Many historians mention bad weather in passing when describing the events of these centuries, but few acknowledge the possibility that the deterioration in climate played a key role in the deterioration of civilization. In his epic *Hutchinson History of the World*, for example, the Oxford historian J. M. Roberts says that the medieval economy was 'never far from collapse', and that agriculture was 'appallingly inefficient'. So when two successive bad harvests in the early fourteenth century reduced the population of Ypres by a tenth, he blames the inefficiency of the infrastructure of society, not the bad weather, seeming to miss the point that in the preceding two centuries, during the little optimum, harvest failures had been rare, and that is why society had thrived and population increased in spite of the 'appalling inefficiency' of the farmers. No historian could fail to notice, however, that something dramatic happened in Europe in the fourteenth century. As Roberts puts it:

> It is very difficult to generalize but about one thing there is no doubt: a great and cumulative setback occurred in the fourteenth century. There was a sudden rise in mortality, not occurring everywhere at the same time, but notable in many places after a series of bad harvests around about 1320. This started a slow decline of population which suddenly became a disaster with the onset of attacks of epidemic disease which are often called by the name of one of them, the "Black Death" of 1348–50.

But why did the harvests fail? Because of bad weather. And why did people succumb to the Black Death (another name for bubonic plague)? Because they were weakened by malnutrition. In the late thirteenth century, life expectancy in England was about 48 years; at the end of the fourteenth century it was less than 40 years. The Black Death was partly to blame, along with other killing epidemics of typhus, smallpox and even influenza (which was still a major killer in the twentieth century; more people died of influenza in a great European epidemic just after the First World War than had been killed on the battlefields of the war itself). In some regions, half the population was wiped out by these plagues. Over the whole of Europe, the population fell by a quarter. Trouble produced more trouble: searching for scapegoats for the disasters, people often took to hunting witches or persecuting Jews; civil unrest brought uprisings in France in 1358 (the *Jacquerie*) and in England in 1381 (the Peasants' Revolt). And in many regions the land became depopulated as people deserted the villages and abandoned their fields.

But which came first, death or desertion? Traditional school history books lay the blame for the depopulation of rural regions on the wave of plagues. But more recently researchers such as Martin Parry, of the University of Birmingham, have pointed out that according to the records there were many villages with uncultivated land in every part of England in the year 1341, at the time of a great survey known as the *Nonarium Inquisitiones*. Farmland was being abandoned, not just in England but across Europe, before the plagues struck. Villages were being deserted before the Black Death did its grisly work on a weakened population, and the villages that were abandoned in the wake of the plague were just those which had been most weakened by the preceding famines, and had already lost two-thirds of their populations before the disease got a grip. Famine and depopulation first, and *then* the ravages of plague, has now clearly been established as the order in which desertion and death altered the face of rural Europe in the fourteenth century; and before the famine had come the worsening of the climate.

Scotland suffered even more than England. The benign climate of the twelfth and thirteenth centuries had allowed agriculture to develop far up the glens, and brought a golden age of civilization in the north, where many English exiles had retreated from the Norman invaders. But in the fourteenth century, crop failures and hunger stimulated clan warfare in the north. In the 1430s, conditions were so bad that bread was made from the bark of trees, since there was no grain; and the civil unrest became so great that in 1436 the king, James I, was murdered while out hunting near Perth. With the king no longer safe in his own lands, the court retreated to the fortified city of Edinburgh, in the south of the country, which became Scotland's capital. In the same decade, further south, the severe weather brought the last recorded evidence of wolves being active in England.

The effects of the climate shift were also felt far away from Europe. In Africa, desert regions became drier and expanded during the fourteenth and fifteenth centuries, while in India various estimates of the total population suggest that a peak of about 250 million was reached in the year 1000, and that as the climate deteriorated this fell to 200 million in 1200 (remember that the characteristic little-ice-age weather drifted westwards from China, and therefore hit India sooner than it did Europe) and 170 million in 1400, with a sharper fall to 130 million by 1550. All this was caused by the development of a stronger, larger circumpolar vortex around the Arctic region, and associated changes in monsoon rains and temperature patterns across the northern hemisphere. Closer to that expanded vortex, in North America, the changes were even more dramatic.

Cowboys and Indians

The Mill Creek people, who lived in the northwest of what is now lowa, were the local representatives of an Amerindian culture that spread across the Great Plains during the little optimum. Between about 900 and 1200, peole lived in permanent settlements in the valleys from the foothills of the Rockies to Colorado, Nebraska and even farther east, growing corn on their farmland and hunting deer in the woods, where oak and cottonwood trees thrived. The period of greatest success for these farmers coincided exactly with the heyday of Viking voyages around the North Atlantic, and it had its roots in the same climatic pattern. With a shrunken circumpolar vortex displaced towards the Pacific side of the Arctic, the prevailing winds around the Arctic came northwards around the Rockies, and were relatively weak. Moist air systems pushing up from the Gulf of Mexico could penetrate far to the north, bringing life-giving rain with them.

But when the circumpolar vortex expanded and pushed south, stronger westerly winds at lower latitudes blocked the path of these moist weather systems up from the south. The westerlies themselves contained very little moisture because they had lost it all in climbing over the Rockies; the plains lie in the 'rain shadow' of the Rockies, and the stronger the westerlies below, the farther east that rain shadow extends. All this is clear from modern studies of weather patterns. Occasional months, or years, when the circumpolar vortex expands and the westerlies are strong can be used as analogues for whole centuries of increased vortex activity long ago.

In the 1960s, climatologist Reid Bryson and anthropologist David Baerreis, both of the University of Wisconsin, decided to investigate how such a strengthening of the vortex affected the farmers who lived on the plains a millennium ago. They knew how the climate around the North Atlantic and in Europe had changed after about 1200, and the impact this had had on Norse and European society. But nobody had investigated how and when any associated climatic changes might have affected the contemporary cultures of North America. In the best scientific tradition, Bryson and Baerreis predicted that the expansion of the circumpolar vortex and the increasing strength of the westerlies must have made the plains dry out in the shadow of the Rockies, corn withering in the fields and the forests drying up and then disappearing. They then set out to obtain archaeological evidence to support their theory, and to find out how the Mill Creek people had adapted to their changing circumstances.

The Mill Creek region was chosen for their study because it was known to be the site of abandoned villages (shades of Europe in the fourteenth century) which had not previously been thoroughly excavated. When the site was excavated by the Wisconsin team in the 1960s, it provided a wealth of information about the way people had lived there, from 900 to 1400. Accurate dates for remains from different layers below the surface were obtained by a standard technique known as radiocarbon dating, in which the residual traces of radioactivity in carbon atoms from the remains of once-living plants or animals are measured. (The technique became headline news in 1988 when it was used to date the Turin Shroud, and proved that it was an artefact less than a thousand years old, and not the burial cloth of Jesus Christ.) For Turin Shroud or Mill Creek bones, the technique is the same. Radiocarbon is produced naturally by the interaction of cosmic rays from space with the atmosphere of the Earth, and so it is present in tiny quantities in the air that we breathe and that plants and animals use as a source of carbon. When the plant or animal dies, the amount of radiocarbon in its remains declines steadily in a known way, as the atoms decay. So measuring the amount of radiocarbon present today tells you how long it is since the plant or animal died.

The Mill Creek people ate a lot of game, and left a lot of bones behind. The remains at one site included bones from bison, deer, elk, fishes, birds and rodents, as well as other unidentifiable fragments. The Wisconsin resarchers looked at the proportions of bones from deer, elk and bison as the centuries passed. In about 900 nearly 70 per cent of the bones were those of deer, just over 20 per cent were elk, and only 10 per cent came from bison. Up to 1100, although the proportion of elk bones declined dramatically while the proportion of bison bones rose, there were still more deer bones than bison bones. But after 1100 there were more bison bones than deer, and by 1200 the remains were 70 per cent bison and 30 per cent deer, with just a few elk bones. At the same time, from 1100 onwards the total number of bones found at the site declined as the years passed.

The significance of this change is clear. Bison eat grass, and deer browse on trees. A prolonged drought that killed trees would have a severe adverse effect on the number of deer, but as the forests declined and the grass spread there would be more plains for the bison to roam. At the beginning of the dry times it was easier to find bison than deer, and the changing diet of the Mill Creek people reflected that. As the drought became worse and the plains became more arid, though, even bison became harder and harder to find. But how did the drought affect the corn? Bryson and Baerreis decided to measure corn production by the number of pieces of broken pottery found in different layers of the soil, corresponding to different ages from the past. Pots were needed both to store grain and to cook it, so the number of potsherds is generally accepted as indicating how much grain there was at the time the pots were made. After about 1200, the number of potsherds found at the Mill Creek site drops dramatically. The total number of bones from elk, deer and bison had already started to fall even before then, around 1100, just at the time when bison became a more important part of the diet than deer. Together, the various pieces of evidence paint a clear picture: during the twelfth century the plains began to dry out, and the lifestyle of the Mill Creek people changed as a result. By 1400 there were no more bones and no more potsherds being left at the Mill Creek site. The villages, like many other villages in the plains, had been abandoned. But, unlike the Norse in Greenland, the Amerindians had not simply been killed off by the bad weather.

Other evidence from many village sites across the plains supports the broad interpretation of a change in the way of life of the plains dwellers, and analysis of remains of pollen in the soil, of example, confirms that forests died out and were replaced by tall grass prairies. Now this, surely, begins to sound more familiar than our earlier story about Amerindian villagers living in settled communities on the Great Plains, farming corn and hunting deer in the woods like Robin Hood. 'Everybody knows' what the plains Indians were like, from the cowboy movies we all watched as children. Plains Indians were nomads, not settled farmers; hunters who lived in tents and followed the great herds of bison (mistakenly called 'buffalo' in most of those movies) as they roamed the grassy plains. That, indeed, was the kind of 'Red Indian' culture that was encountered by Europeans who moved west into the plains in the eighteenth century, and the open prairie provided the range land needed for cattle ranching on a large scale, which in turn provided work for the tough cattle-hands of the old West, who became the stereotypical cowboys who shared those movies with the stereotypical Red Indians. Cowboys and Indians, as portrayed in the western movies, owed their existence to the drying out of the Great Plains as the circumpolar vortex expanded and the world slid into a little ice age.

What has happened in the past can happen again. In the twentieth century, as we shall see, the world warmed again, and the circumpolar vortex contracted. Rain returned to the plains, not in enormous quantities, but enough with the aid of modern agricultural techniques to put the soil under the plough and to see corn grow there again in quantities that the Mill Creek people would never have thought possible. But even in the more benign climate of the twentieth century, there has been only just enough rain, and farming on the Great Plains rests on a knife edge. The dustbowl years of the 1930s, and even the drought of 1988, show what a small fluctuation in climate can do to grain production in the region, which has been so successful in most years recently that it is known as the 'breadbasket of the world'. An underlying theme of Bryson's work has been his concern that the circumpolar vortex might expand once again, bringing prolonged drought back to the region, and ripping the heart out of American farming. But it is possible to fall either way off a knife edge. In the past couple of years other researchers have expressed more concern about another possibility, that the world may now be warming so much that the plains are simply becoming too hot for farming to continue as before, with the grain-growing region shifting northwards into Canada.

Either way, there is a lesson to be drawn from the way Mill Creek farmers became nomadic Red Indians. Apart from the cold in Greenland, the farmers were faced with exactly the same choice as their Norse counterparts. Farming in settled communities became impossible, but hunting, either on the plains or on the sea and ice, remained a viable way of life. The Norse in Greenland stuck to their farming, and died. The Mill Creek people literally upped sticks and rode off after the game, and survived. Their life *style* ended, but their life continued. Whichever way the climate of the world jumps in the twenty-first century – back into a little ice age or on into hothouse warmth – our global society will have to adapt if it is to survive intact. The rigours faced by our ancestors from the sixteenth to the eighteenth centuries, the deepest trough of the little ice age, show just how successfully human society ought to be able to cope even with the worst the weather can do.

When Dick the Shepherd blows his nail

We hope you are suitably impressed by our literary allusion. It comes from William Shakespeare's *Love's Labour's Lost.* Shakespeare, who lived in the late sixteenth century in the southern part of England, wrote of a time:

> When icicles hang by the wall,
> And Dick, the shepherd, blows his nail,
> And Tom bears logs into the hall,
> And milk comes frozen home in pail

But why did Shakespeare write of milk coming home frozen in the pail? Not just because it made a good poetic image, nor simply that in his day cowsheds and milking parlours were unheated. The milk froze because winters in Shakespeare's day were significantly colder than the winters we have been used to in the second half of the twentieth century.

The thermometer had not been invented then, and there were no convenient glaciers in southern England where oxygen isotopes could record the temperature. But there is another useful indicator of the severity of sixteenth-century English winters, not far from Shakespeare's door. The River Thames provides a marker for severe winters down the centuries. In the worst winters, the river froze into a solid surface thick enough for people to walk on, drive carriages across, and even build tented cities known as Frost Fairs, and all these events are recorded in the history books. The famous old London Bridge referred to in the nursery rhyme had been built in 1209, and helped the freezing process. The pillars of the arches of the bridge were so thick that there was more pillar than there was gap in between, and water piled up above the bridge, rushing through the arches as if over a weir. Debris such as tree branches (or chunks of floating ice from upstream) also piled up against the bridge, and in cold weather a sheet of ice would form at this obstruction first, and grow steadily upstream. But the bridge was not the sole cause of those great Thames freezes: in 1269/70 the river froze so far *down*stream that goods had to be sent overland to London from the Channel Ports, instead of coming upriver by boat in the usual way. It also froze in 1281/82 and in 1309/10. But the little ice age really began to live up to its name in London in the fifteenth century. Between 1407/08 and 1564/65 the river froze six times. Horses and carts were driven across the frozen river on several occasions, avoiding the toll collectors on the bridge. Henry VIII drove a carriage on the river on at least one occasion, either in 1536/37 or 1537/38, and Queen Elizabeth I used to take a regular daily stroll on the ice in the winter of 1564/65 (this was the first winter after Shakespeare's birth, although he can hardly have remembered much about it).

In the seventeenth century, the river became an almost regular winter-sports attraction. The first recorded Frost Fair was held on the Thames in 1607/08. Booths set up on the ice sold food, beer and wine; there were bowling, shooting and dancing on the ice by way of entertainment. Another Frost Fair was held in 1620/21, and according to legend the sport of skating was introduced to Britain in 1662/63, when the King, Charles II, watched the skaters on the frozen Thames. But the greatest Frost Fair of all was held in the worst winter on record, in 1683/84. This was 470 years after the old London Bridge was built, and it was still standing. Only a little over three hundred years separates us from the greatest of Frost Fairs. To the people of 1684 the building of the bridge was ancient history, more remote from them in time than they are from us. But few people in London can have thought much about ancient history that winter.

So many shops and booths appeared on the ice that it was like another city. They were arranged in proper streets, and one enterprising printer made a small fortune by printing peoples' names on slips of paper carrying the legend 'printed on the Thames'. Even the King (still Charles II) and his family had their names printed when they visited the tent city on the ice between the banks of the Thames.

It is recorded that the ice was 11 inches (28 cm) thick in places, and the river was completely frozen for two months. But the frozen river was not the only marker of severe cold that winter. Where the ground was free from snow cover, in the southwest of England, the soil was frozen to a depth of more than a metre, while the sea itself froze in a coastal band of ice 5 km wide along the Channel shores of England and France. At the coast of the Netherlands the ice sheet was more than 30 km across, and no shipping could get through. These were all features of a winter so severe across England (and Europe) that it has left its own mark in literature, forming the basis for the fictional winter blizzards in the novel *Lorna Doone*. A hundred years after Shakespeare's time, the cold weather still provided good material for storytellers. Altogether, the river froze ten times in the seventeenth century, and a further ten times between 1708/09 and 1813/14, but it has never frozen since. That is partly because the old bridge was demolished in 1831 (it was, after all, falling down by then, as the nursery rhyme tells us), allowing water to flow more freely; it was partly because industrial pollution and the heat of waste water made it harder to freeze anyway; but it was chiefly because the world got a little warmer after about 1850.

Shakespeare's frozen milk and the Frost Fairs on the Thames were as much a part of the little ice age as the emergence of bison-hunting as a way of life for plains Indians. The thermometer had been invented by the end of the seventeenth century, and early records indicate that during the 1690s temperatures in southern England were 1.5°C lower than in the middle of the twentieth century. Between 1650 and 1800 Europe was colder than it had been at any other time since the last ice age ended, and the cold, though not everywhere quite as severe, embraced the whole world, the first time since the last ice age that every region had cooled at the same time. Such a severe deterioration in climate could not fail to leave a mark on society. But there are so many marks, in so many places, that it is difficult to know which ones to single out for mention. Our potted history of the freezing of the Thames, for example, includes a mention of both Elizabeth I and Charles II. In between, in the middle of the seventeenth century and in the worst of the litte ice age, England experienced civil war and first the abolition and then the restoration (in a more limited form) of the monarchy. There were many other reasons for the turmoil of the times, and we would not be so foolish as to suggest that privations caused by crop failures and bad weather were the only reasons for the success of Oliver Cromwell's revolution; but it would be equally foolish for historians to ignore the fact that one reason why people were cold, hungry and desperate for political change in the seventeenth century was because the little ice age was at its worst. Other political upheavals, some of which still affect society today, also have their roots in the little ice age. We give here a few examples; you can undoubtedly find more in the history books (indeed, if you are like us then no history book will ever seem quite the same once you know how the climate changed in historical times).

Out of the ice

The cold of the little ice age extended around the world. For the sixteenth century there are good records of the weather from many parts of Europe. The pioneering astronomer Tycho Brahe, for example, kept a weather log in Denmark in the late sixteenth century, and in England diarists such as John Evelyn and naturalists such as Gilbert White kept records in the seventeenth century. Further north, the Arctic sea ice spread southwards. In the 1580s, the Denmark Strait between Iceland and Greenland was blocked by pack ice in summer on several occasions. In the worst year for pack ice on record, 1695, Iceland itself was completely surrounded for 12 months, with no ships able to enter or leave the ports; and in 1756 the coast of Iceland was beset by ice for 30 weeks. Just at this time, between 1690 and 1728, there are several reports of lone Inuits arriving in the Orkney Islands, north of Scotland, in their kayaks; and on at least one occasion an Inuit in a kayak reached the mainland, near Aberdeen. Not only had European *H. sapiens* made the journey across the Atlantic to link up with North America *H. sapiens*, but the eastward spread of the species around the globe, out of Africa and across Asia to the Bering Strait, Canada and the high Arctic, had very nearly come full circle.

Scotland, as this evidence implies, cooled even more during the little ice age than regions further south. The sea to the north of the country seems to have been 5°C colder than in the middle of the twentieth century, and there was permanent snow on the tops of Scottish mountains through the seventeenth and eighteenth centuries. There were repeated famines and a dramatic depopulation of the country as a result. Many emigrants became mercenary soldiers, fighting in the many wars that troubled Europe at the time (and which themselves had their roots, at least partially, in the deterioration in the climate). In the later part of the little ice age, many emigrants went to America. But the emigration that still has unwelcome repercussions today came in 1612, when King James VI of Scotland (by then also James I of England, and thereby ruler of Ireland) forcibly evicted the Irish from the northern province of Ulster, in Ireland, and relocated Scottish farmers there instead. His motives were partly to increase his influence and build up a power base in Ireland, but also to provide a haven for refugee Scots. In some ways the scheme was a success, for by 1691 there were 100,000 Scots in Ulster, with only ten times as many left in Scotland, and more immigrants followed after another wave of cold and famine in the 1690s. But in the long run the move was a disaster. These transplanted Scots were the ancestors of the modern Protestant population of Ulster, and the root cause of the troubles that continue in Ireland to the present day.

But for Scotland, this was far from being the end of the tale of climatic woe. Between 1693 and 1700 the harvests failed in seven years out of eight. About half the population died – more, in many places, than had succumbed to the Black Death – and the country was forced by circumstances into a full union with England in 1707 as a means of ensuring aid from the south. It was just as bad on the other side of the North Sea. In the last two decades of the seventeenth century, glaciers overran farmland in Norway and the harvests failed three times in a row on two separate occasions, in 1685-87 and in 1965-97.

Farther afield, in parts of Africa, it was flood, not frost, that was the problem. As the circumpolar vortex of air expanded, the rain belts north of the equator were squeezed back towards the south. That reduced the rainfall on the fringes of the Sahara, allowing the desert to expand, but increased the rainfall farther south and produced great floods in, for example, the region of Timbuktu. Lake Chad, between 13°N and 14°N, was 4 m higher than today in the seventeenth century, and there were heavy summer rains over Ethiopia, bringing very high floods in the River Nile. In India, however, the expansion of the vortex pushed monsoon rain systems aside on many occasions in the seventeenth century, bringing drought more frequently than in the twentieth century.

North America, by now beginning to be settled by Europeans, also suffered. The winter of 1607/08 was particularly severe, spreading death among both the European and the native populations of Maine, and leaving Lake Superior frozen so hard that around the edges of the lake the ice could still bear the weight of a man as late as June. But the presence of European settlers in North America is a reminder that the little ice age also coincided with a great expansion of European culture and a movement of both people and that culture out of the European heartland. The expansion was very much due to the invention of better ships and improved methods of navigation, and in some ways it was in spite of the severity of the weather, showing how effective even what we would regard as primitive technology can be in combating the climate. But as far as the movement of people was concerned, the European expansion was also partly caused by the shift in the climate. Scots did not just settle in Ulster and roam Europe as mercenaries, they also moved in large numbers to North America, along with other people seeking a better life in a new land.

Europeans had to move out across the ocean if their culture was to develop and spread. To the south and east they were hemmed in by the Islamic Ottoman Empire (which had no need of great ocean-going ships, since the empire stretched from Spain to India and could be navigated almost entirely overland). To the north and east, Russia provided an equally effective barrier against European expansion. New lands, trade routes, conquest and expansion were to be found only by sea. Columbus crossed the Atlantic in 1492. Within a hundred years, a combination of the availability of new lands and the pressure on poor people in Europe resulting from the growing severity of the little ice age began to make emigration attractive to many. In the seventeenth century about a quarter of a million British emigrants went to the new world; in the eighteenth century, 1.5 million made the trip. Set these figures against the estimated population for England of just 6 million in 1700, and the importance of the new lands in providing a safety net for people who might otherwise have starved in the worst decades of the little ice age is plain. Even with this safety net, there were famines, civil war and a return of plague (in 1665) to trouble England alone, with similar difficulties across Europe; what might have happened to European society if there had been not safety net? There were even 200,000 Germans in North America by 1800, although Germany had no colonies of its own, and well over 2 million Europeans in all had settled north of the Rio Grande by then. South of the river, only 100,000 Spaniards and Portuguese had left their homelands to settle in the Americas at that time, even though the two great nations (unlike Germany, for instance) had dominated the exploration of the new world in the early decades, and had been (and remained) major sea powers. It is hard to explain, on the face of things, why Germany, with no colonies, had 200,000 emigrants to North America, while Spain and Portugal, great colonial powers both, could muster only 100,000 emigrants to Central and South America between them. But Spain and Portugal, of course, lie farther south than England and Germany, and suffered correspondingly less badly during the worst ravages of the little ice age. There was less incentive to emigrate as long as life was still tolerable at home. Their migrants went to rule the colonies; German migrants (and British, French and other more northerly European people) went to work the land themselves and make a home in America.

Ironically, the end of the little ice age brought one of the greatest mass migrations westward from Europe. In 1846 the summer was warm in Europe generally, and wet in Ireland in particular. Conditions were ideal for the spread of the potato blight fungus. Once it got a grip there were continuous outbreaks for six consecutive years, with the effects on the starving population made worse by an outbreak of typhus. In 1845 the population of Ireland had been 8.5 million; a million died in the famine, and many more emigrated. By 1900 the population was down to about 4 million, and it has never recovered to the pre-famine levels. Coincidentally (?), the 1840s were a time of trouble in Europe generally, culminating in the 'year of revolutions', 1848. But that is another story. For our purposes, the Irish potato famine stands as the archetypal example of how climatic events forced mass migrations from Europe, although the kind of climatic shifts responsible were different in the 1840s from those of the preceding two or three centuries.

There is no doubting the harshness of the sixteenth and seventeenth centuries. Historians have long puzzled over the underlying causes of a dramatic period of economic inflation that afflicted Europe after about 1500. Prices rose fourfold in a century, not much by some modern standards but unprecedented at the time. The prices of agricultural produce rose particularly sharply, but wages did not keep pace, and the poor starved. There were rebellions, uprisings and what Roberts calls 'a running disorder' which 'reveal both the incomprehensibility and the severity of what was going on'. Why did it happen? Was the population increasing too rapidly for new farming techniques to keep up? Were the difficulties caused by a flood of gold from the New World? Or was it, a point few historians seem to bother to argue, because of the deterioration in the climate? We believe that climate played a bigger part than most people have yet acknowledged in the movement of people out of Europe over the past 400 years or so. Just as people are children of the ice age, the Europeanization of the world is, at least in part, a product of the little ice age. The corollary of that interpretation of history is that climate may also play a more important part than most people realize in determining the future of human society. But before we move on to that, we would like to close our account of the little ice age with some examples of how even the last and relatively weak waves of cold coloured the lives of those who lived through them, and have left images that linger in our minds today.

What the Dickens?

Back in the sixteenth century, the beginning of the worst phase of the little ice age helped to create a whole new school of painting. The winter of 1564/65, which we have already mentioned, was the worst since the 1430s. In February 1565, in the very depths of the winter, Pieter Bruegel the Elder painted the picture that started a trend. It was called 'Hunters in the Snow', and it depicted just that. Lamb says that this was 'the first time that the landscape itself, albeit in this case an imaginary landscape, has been, at least in essence, the subject of the picture rather than the background to some other interest'. Before that harsh winter, in 1563, Bruegel had painted a stereotypical picture of the Nativity of Christ, with the Three Kings offering homage to the infant Jesus, the main focus of the scene. After that winter, in 1567, he painted the Nativity again, but this time what we see is a snowscape, a picture of people struggling to cope with harsh weather conditions, in which the presence of the baby Jesus is not even obvious to the casual observer. In the decades that followed there were to be many more such scenes portrayed on canvas, by Bruegel and others.

That was at the beginning of the most severe part of the little ice age. Near the end of the downturn in climate, after a brief amelioration in the late eighteenth century there was more harsh weather to come in the first decades of the nineteenth century, not as bad as the 1690s, but bad enough for people who had started to get used to something rather better. Fashions changed to match the weather. In the 1790s the new ladies' fashion from France exposed a large part of the upper body; by 1820, fashionable dress was much more modest, not because of moral objections to the new fashion but because the cold weather made it impractical. One reason for the change in the weather may have been an upsurge in volcanic activity, with several major eruptions around the world. Dust and other pollution from explosive vocanic eruptions penetrates high into the stratosphere, forming a haze which blocks off part of the Sun's heat, sometimes for years after a big eruption. In 1815, one of the most spectacular of all eruptions, at Tambora in the East Indies, threw at least 15 km^3 of material into the upper atmosphere, producing spectacular sunsets around the world and making 1816 so cold that it has gone down in history as 'the year without a summer'.

During the cold decades at the start of the nineteenth century, the greatest of all British painters, J. M. W. Turner, began to depict the spectacular sunset effects which became his trademark. They were based on reality, not simply a product of his imagination. Historical climatologists can even use paintings from different periods to help them reconstruct the changing weather patterns of the past. Pictures of representational outdoor scenes, typically painted in summer in Europe, show an average of 80 per cent cloud cover in the period from 1550 to 1700, between 50 and 75 per cent cover in different decades of the eighteenth century, and about 75 per cent cover in the period when Turner and John Constable were active in

England, from about 1790 to 1840. Twentieth-century artists show on average about 60 per cent cloud cover.

And if Shakespeare's imagery was influenced by the weather of his day, so too was that of Charles Dickens. Dickens was born at the tag end of the last phase of the final cold spell of the little ice age. In the first nine years of his life, from 1812 to 1820, there were six white Christmases in London, either because snow had fallen or because of heavy frost. They made a deep impression on the boy, who grew up to write stories which contain the definitive descriptions of traditional white Christmases in southern England. Even by the time those stories got in to print, however, white Christmases in London were a rarity, and today they are almost entirely a thing of the past. But Dickens gave the Victorians their picture of what Christmas 'ought' to be like, and that picture is still preserved on every Christmas card that has a coach-and-four hastening through the snow, or a robin sitting on a frosty bough. Our Christmas cards owe more to Dickens's boyhood, and the last gasp of the little ice age, than to the climate of the twentieth century, but a milkmaid from Shakespeare's day would have found the scenes familiar. Will the snow and ice return? Or has the little ice age left us for good? Before we bring our tale to a conclusion, we promise to take a look at how the activities of the brainy biped, a product of the rhythms of the ice epoch, may now be making snow at Christmas, or any other time of year, a thing of the past, not just in London but everywhere except Antarctica. But first, having brought the story of human evolution and the influence of environmental changes on our evolution up to date, the time has come to look at how our understanding of animal behaviour in general can provide insights into what being human is all about. For, remember, although that crucial 1 per cent difference does make us special, our 99 per cent chimpanzee inheritance does mean that many of the rules that determine the behaviour of animals in general, and primates in particular, also apply to us. Many people are horrified at the thought, but we find it both intriguing and hopeful, since the better we understand the reasons for our own behaviour the more chance there is of making our animal instincts work beneficially in the context of civilization. The science behind all this goes by the name of sociobiology.

CHAPTER SEVEN

In Praise of Sociobiology

Sociobiology is the study of all forms of social behaviour in all animals, including humans. That sounds innocuous enough, but sociobiology naturally includes the genetic bases of behaviour. A hyena has a very different social life from an albatross, say, and this is fundamentally because the DNA inside the cells of the two species codes for different things. The fact that an albatross has two wings and two legs, and is covered with feathers, while a hyena has four legs and no wings, and is covered with hair, is due to their different genetic inheritance. Clearly, a very large part of sociobiology is about genetic inheritance, which predisposes individual members of different species to do certain things well, and other things badly or not at all. Of course, the more interesting aspects of this study concern differences more subtle than those between wings and legs. Why do lions, for example, find it advantageous, in terms of evolutionary success, to form groups (prides) in which one or two males dominate and mate with all the females, while the rest of the males get no chance to reproduce? Why do most birds spend so much time and effort raising their young, while most frogs simply abandon their eggs to take their chance in the world? And so on. The principle is still the same: a bird behaves like a bird, and a frog like a frog, at least partly because of their genetic inheritance.

When this straightforward line of biological and evolutionary reasoning is applied to humans, it causes such a strong reaction in some quarters that the very term 'sociobiology' has become almost a dirty word to a small minority of biologists. This reaction does not come from those who believe in the literal word of the Creation story told in the Bible, and therefore cannot accept that people should be regarded by science in the same way as other animals. Ironically, the vehement opposition to human sociobiology has come from the other end of the religious spectrum, from self-acknowledged Marxists and atheists who mistakenly believe that sociobiologists are claiming all human behaviour to be so rigidly programmed by our genes that there is little or no scope left for free will.

We would never have dreamed that sociobiology needed singling out for praise (or justification) in this book, had it not been for a lunchtime discussion we attended at London's Institute of Contemporary Arts in 1984. The theme of the discussion was the role of men and women in modern society, and one of us argued that in strict biological terms the human species no longer 'needs' sex, and that both sexual reproduction and males are redundant hangovers from our evolutionary past (more of this later). This drew an almost apoplectic response from one of the other people present, who protested, 'But that's blatant sociobiology!' Somewhat baffled, we agreed. Of course it was, and is, sociobiology. How else do you study the evolutionary significance of such things as the existence of two sexes in mammals? Our 'opponent' sat back, satisfied. In his eyes we were condemned out of our own mouths, as sociobiologists, and were therefore beyond the pale. Nothing further needed to be said on the matter, and his mind was closed to further discussion.

So, after the meeting we thought we had better find out what all the fuss was about. Why was it that the simple idea of applying sociobiology to humans, which seemed so straightforward to someone coming to the study of evolution from a training in the physical sciences, should arouse such passions and be dismissed so strongly, even if those passions burned only in a minority of breasts? We soon found that the place to start investigating the phenomenon was with the work of Edward Wilson, of Harvard University, who published a landmark book called *Sociobiology* in 1975.

What is sociobiology?

Wilson did not invent the kind of study now called sociobiology, although he was the first to use the word with its present meaning. The roots of investigations into the genetic bases of the behaviour of animals (including humans) go back a long way, and some of the key developments in the early 1970s came from Robert Trivers, who is now at the University of California at Santa Cruz. Wilson's best-selling book caused debate and aroused passions (and became a best-seller) because he said specifically that, in scientific investigations of behaviour, people should be treated in the same way as other animals. To some extent, the behaviour of animals is governed by their genes. People are animals. Therefore, to some extent, the behaviour of people is governed by their genes. The logic seems impeccable, especially when you recall that some 99 per cent of our own material is identical to that of the pygmy chimpanzee. The evolutionary rules that apply to chimps apply to us as well. But still, some people are uncomfortable with that logic, just as many Victorians were unhappy with Charles Darwin's claim that there was *any* biological relationship between mankind and apekind.

In recent years two extreme views have been put forward to explain human behaviour, and much of human history. Konrad Lorenz and Robert Ardrey espoused the view that human beings are driven by innate aggression, genetically determined, which has moulded the development of human society through repeated warfare and which finds an outlet today in activities such as football hooliganism. Ardrey in particular, in a series of best-selling books, promoted the idea that human violence is an inevitable fact of life. B. F. Skinner, on the other hand, postulated that each newborn human infant is a blank slate, capable of being moulded in any direction depending on the kind of stimuli it receives from the world around it. Train the child to be a warlike aggressor, and you will get an aggressive adult; train the child to be a peaceful farmer, and you will get a nonviolent adult agriculturalist. As with most extreme views of the world, the truth lies somewhere in between. People are born with certain innate abilities and inclinations, but they are also moulded by their cultural surroundings. We are born with a predisposition to speak, for example, but the language we speak as adults depends on what we hear when we are young.

As we have seen, flexibility, linked with our large brain and long childhood, is a key human attribute. Lorenz himself argued that although aggressive behaviour in human has an innate basis, society might be changed to cope with these basic drives and channel aggression into socially useful forms of behaviour (a different view from Skinner's suggestion that the child might be changed to fit society); some of his 'followers', however, have rather ignored this point. It is interesting to investigate how much the adult behaviour is a result of inheritance (nature) and how much to upbringing (nurture), but always remember that the ability to be flexible and adapt is itself something that is part of our genetic inheritance, and that it has evolved because it is successful. During natural selection, *any* device which helps to ensure that a higher proportion of certain genes is passed on to the next generation will become a characteristic feature of the species that carries those genes. That is why (within limits) people have the ability to learn new tricks and adapt to different cultural environments.

If culture were the overriding influence on being human, then surely it would be possible to raise a pygmy chimp from birth in a human household, and end up with a passable imitation of a human being. In fact, the genetic difference of only 1 per cent makes this impossible. Humankind is indeed a cultural animal, but our cultural nature and our instincts alike are themselves genetically determined, like our naked skin and our large brain.

The extreme opponents of human sociobiology are concerned that any evidence of a genetic basis for human behaviour will strengthen the case of the Ardreys of this world, who argue that human aggression is innate and cannot be controlled. Such biological determinism in its extreme form can be, and has been, used as a justification for war, for the domination of society by one class, for the subordinate position of women in society and for many other evils. We agree wholeheartedly with the opponents of sociobiology that these are indeed evils which ought to be eradicated from human society. But they are missing the point: sociobiology does *not* presume that our fate is determined by biology in this way.* Culture *is* a big

* If our behaviour were indeed determined 100 per cent by these kinds of biological imperative, it would of course never occur to us to question them or to regard aggression, rape, racism and the rest as objectionable. They would be perfectly natural to us. The fact that we question such activities is itself conclusive

influence on human behaviour (thanks to our genes), and we have the intelligence (thanks to evolution) to analyse situations and act on the basis of reasoned argument, instead of by instinct. People are rather unusual African apes, and our unusual attributes have to be taken into account. Indeed, what matters is that we should try, through sociobiology, to understand what our animal inheritance predisposes us for, so that we can decide whether that predisposition is good or bad, and can take suitable steps to overcome it where necessary. If, for instance, people are innately somewhat aggressive and suspicious of foreigners (possibilities we touch on later in this book) it should be advantageous to understand that and to ensure that we use our intelligence to avoid conflicts.

It seems to us that the position has been best summed up by Wilson himself, writing in the *New Scientist* in 1976:

> A crucial issue with which sociobiology has to grapple [is]: what are the *relative* contributions to human behaviour of genetic endowment and environmental experience? . . . we are dealing with a genetically inherited array of possibilities, some of which are shared with other animals, some not, which are then expressed to different degrees depending on environment.* (our italics)

It is hard to see how any evolutionary biologist can take exception to such a statement. It is true, unfortunately, that extreme right-wing groups have misued some of the ideas of sociobiology as 'justification' for racism, claiming that biology 'proves' some people to be inferior to others. What seems to have happened is that extreme left-wing groups have mistaken this abuse of sociobiology by their bitterest foes for the real thing, and have not looked to see what researchers such as Wilson and Trivers are really saying. Discarding sociobiology because a few neo-Nazis claim it to imply that certain races are inferior is like banning the manufacture of knives because a few people commit murder with them. What we mean by sociobiology is encapsulated in the passage quoted from Wilson's *New Scientist* article. And a great deal has happened since 1976 to put sociobiology in general, and human sociobiology in

proof that we are not blind victims of genes 'for' aggression, rape, racism and the rest.

* 13 May 1976, p. 344.

particular, on a footing so secure that no blasts of hot air can shake it.

John Krebs, of the University of Oxford, has summed up the major themes that have emerged as sociobiology developed in the ten years following the publication of Wilson's book.* The most important development is the introduction of games theory into the study of evolution, with the idea of the evolutionary stable strategies, or ESS.

Strategies for success

Evolution proceeds because there is variety among the individuals of a species, and because there is selection of the individuals best fitted to their environment. The ones that are well adapted to their environment do best and leave more offspring than the less well fitted, so their genes spread. The competition in Darwinian evolution is not so much between different species, but between individual members of the same species. And John Maynard Smith, who coined the term 'evolutionarily stable strategy', has given us a mathematical basis for understanding the individual strategies for survival.

It is based on a branch of mathematics called games theory, which has been developed in sound mathematical detail in recent decades, originally because of its importance in 'modelling' different aspects of war and calculating the odds of success for different strategies in different situations. It carries across almost unchanged into the competition between individual members of a species to make the best use of the resources available in the ecological niche occupied by that species. The best way to give an idea of what games theory and ESS is all about is by an example, from Maynard Smith's work. This is the scenario of 'Hawks *v.* Doves'.

Imagine, says Maynard Smith, a population of animals all of the same species. Each individual member of the population may behave either as a Hawk, denoting aggressive behaviour, or as a Dove, denoting peaceful behaviour. (In fact, doves are pretty aggressive birds, but they have come to symbolize peace

* In the anniversary article 'Sociobiology ten years on', *New Scientist*, 3 October 1985, p. 40.

in a metaphorical sense.) When a Hawk finds a piece of food, and another member of the species is present, it will always fight, if necessary, in an attempt to get the food. When a Dove finds a piece of food, but another member of the species is present, it will run away at once if attacked. If it meets another Dove, it will try to bluff by making a threatening display, but will eventually leave the food and retreat. These are the basic rules programmed into Maynard Smith's mathematical model. The question he then asked is, 'what proportion of Hawks and Doves represents a stable state of the population?'

This is a purely hypothetical example, so let us put some purely hypothetical (but sensible) numbers in, to indicate the value of the food the animals need. If the individual gets to eat the food, it scores 50 points. If it runs away, of course, it scores 0 points. If it gets in a fight and loses, it will be injured, and scores –100; and if it gets involved in a mutual-threat display before running away, it scores –10 for the waste of time. The numbers are arbitrary, but they demonstrate the relative status of each possible outcome; these particular numbers are taken from Richard Dawkins's book *The Selfish Gene*. The most successful individuals are those that eat the most food and avoid getting hurt, so their genes will be passed on to the next generation. The points are equated with reproductive success, and ESS theory addresses the question of whether there is a stable mixture of Hawks and Doves in the population which will persist from one generation to the next.

The first thing that comes out of the calculation is that neither a population of all Doves nor a population of all Hawks is stable, in this sense. Consider Doves first. If everyone is a Dove, then every time there is conflict both individuals produce a threat display which costs –10 points. But one of them runs away first, so the other one gets the food and scores 50 points, giving it a net gain of 40+. The average 'score' is 15, obtained by taking 40 and –10, adding them together and dividing by 2. So far so good. Everyone gets to eat, nobody gets hurt, and the all-Dove society looks healthy. (You can also see from this very simple example why many species of animal have evolved very elaborate and long-lasting threat displays, attempts to stare the opposition down without actually getting in to a fight. People, like other animals, often do the equivalent.)

But now consider what happens if a Hawk appears in the population (by mutation). It wastes no time with threats, it just chases away the opposition and wins the prize every time, scoring 50 points. As long as Hawks are rare, so that they seldom meet one another and fight, they will do very well indeed, scoring 50 points where the Doves are only scoring 15 on average, and so spreading their genes throughout the population.

So what happens at the other end of the scale? In an all-Hawk society, every meeting produces a bitter conflict. The winner scores 50 points, the loser scores −100 and the average is a pathetic −25. (Such a population would survive, of course, only so long as conflicts were rare and most food could be picked up without meeting another Hawk and getting in a fight.) But what happens if (by mutation) a Dove appears? By running away every time a Hawk threatens, and scoring 0 points in every conflict, but picking up food when nobody else is around, the Dove does relatively well. So Dove genes spread, up to a point.

Clearly, there must be a stable state somewhere between these extremes where the population of Hawks and Doves stays the same, and both strategies provide the same average gain in any conflict. For the particular numbers we have chosen here, the stable population contains five-twelfths Doves and seven-twelfths Hawks – seven hawks for every five Doves – and each individual scores 6.25 in each conflict, on average. In terms of real populations of real animals, we can recast this slightly. It is also a stable strategy for *each individual* to act like a Hawk seven-twelfths of the time and to act like a Dove five-twelfths of the time. Genes which operate on the body as if they were saying, 'be aggressive just over half the time, but be a pacifist a bit less than half the time, and select which you are going to be at random in any one conflict' will be the most successful genes. The ESS can be either for each individual to play Dove x per cent of the time, or for x per cent of the individuals to play Dove all the time.

However, there is something peculiar, and significant, about the numbers we have come up with. The ESS provides each individual with 6.25 points per conflict. The all-Dove scenario gave each individual 15 points, more than twice as much. Every individual in the population would be better off if they were all doves: 'for the good of the species', the all-Dove scenario is better than the ESS! But evolution does not work like that. It operates on individuals, not species, and the stable society is one in which, it turns out, every individual does worse than he or she could. This clearly has implications for human behaviour. It is a concept so basic and informative that it underpins most of the rest of this book.

Sociobiology today

Two themes related to each other are the puzzle of why individuals should so often be willing to cooperate with one another – as in the cooperative social life of a pride of lions – and why there is competition and disharmony even within such cooperative groups, as when the dominant male is supplanted by a younger rival, who promptly kills all the infant cubs in the pride. The roots of this kind of behaviour, both cooperation and competition, can be found in the concept of the 'selfish gene', the argument that the actual units of natural selection are the genes themselves, even though selection has, perforce, to operate (like the genes) on whole bodies.

This idea is important enough, especially when linked with the ESS, to merit a little further discussion. It was presented most forcefully by Richard Dawkins in his popular book *The Selfish Gene*, and he has developed the theme further in his even better book *The Extended Phenotype*, where he replies to earlier criticism and, in our view, thoroughly confounds the arguments of his critics, while developing the concept of the selfish gene more completely and more clearly. Dawkins argues that bodies – phenotypes – exist primarily because they are the means by which genes ensure their own replication, and should be viewed as such if we are to understand the workings of evolution. In the words of the biological aphorism, a hen is simply the egg's way of making more eggs. Or, in human terms, a person is simply the genes' way of manufacturing more copies of themselves.

Life on Earth is all about the replication – copying – of strands of DNA. The complexity of life we see around us, including our own species, is a result of competition and selection which has produced a variety of different ways of ensuring the copying of DNA, and has produced many kinds of DNA, many genes, along the way. This is a blind process operating in accordance with the laws of physics and chemistry, and statistics, with no guiding intelligence behind it. But with that clearly understood, it is often convenient to anthropomorphize somewhat, and use everyday expressions to discuss how the genes ensure their replication. We may say, for example, that a gene 'wants' to ensure that copies of itself get spread among subsequent generations. Obviously one effective way to do this is to 'help' the body in which it lives to reproduce and leave many offspring behind. That is basic Darwinian evolution; genes that make a body better fitted to its environment will inevitably be selected for and will spread. But this is not the fully story. Because of the way genes are inherited, half from each parent, there is a 50–50 chance that two offspring who share the same two parents will have any one of their genes in common. On average, half the genes of full siblings will be identical in each body, each phenotype. Or, from the 'point of view' of a particular gene residing in the cells of one body, there is a 50–50 chance that the cells of the body of a sibling carry copies of that same gene.* So, if one sibling helps another to find food, or a mate, and, ultimately, to reproduce, then he or she is helping to ensure the spread of very many of the genes that are in his or her own body. If a gene arises which causes the body it inhabits to behave in such a way that the survival of siblings is encouraged – even if this is merely a side effect of whatever influence the gene has on the phenotype – then the gene will spread, because half the siblings carry the same gene. (Assuming, for the sake of simplicity, that the gene has no harmful effect on its own phenotype. The mathematics of the

* We use the term 'gene' a little loosely here, where our more rigorous biological friends would use the term 'allele'; we believe our meaning is plain, and hope they will forgive us. Perhaps we should also clear up another potential source of confusion. Humans and chimps share 99 per cent of their DNA, in the sense that the human 'blueprint' (or recipe) is 99 per cent the same as that of the chimp 'blueprint' (or recipe); but two siblings of either species may share 50 per cent of their DNA, in the sense that they have that many identical alleles. The rest of the DNA in each individual is still the kind appropriate for its own species. Two sisters, one with blue eyes and one with brown, are still members of the same species; a man and a chimp, both with brown eyes, are members of different species.

ESS can handle such subtleties, though: in the simple case of the benefit to siblings being more than twice the handicap resulting to the carrier of the gene, then the gene ought to spread, since with a 50–50 chance of the gene being present in the siblings the benefits *to copies of the gene* will then outweigh the handicap *to one individual's phenotype*.) At a stroke, this concept explains, in a qualitative way, why close relations should cooperate with one another, and why members of a pride of lions will, in general, be relatives. The mathematics of population genetics can put numbers into the calculations, and when this is done the combination of the idea of selfish genes and games theory can very often account for otherwise puzzling biological phenomena.

One very simple example demonstrates the power of this approach and shows how sociobiology has stood the test of time. This kind of cooperative behaviour would be useless if individuals had no way of recognizing their close relations. So it is a requirement of sociobiology that close relations, especially full siblings, should be able to identify each other. This is not something that emerges naturally from any other version of evolutionary ideas, and it was not anticipated, even as recently as 1970. But studies by Paul Sherman, of Cornell University, of the behaviour of a creature known as Belding's ground squirrel have shown that these animals distinguish between litter-mates that are full siblings and those that are half-siblings, having the same mother but a different father. The ground squirrels behave differently towards each type of relation: their social life is moulded by biology. It does not matter how the distinction is made (though we might guess that smell has something to do with it). What matters is that it occurs at all.

Of course, this is an example of the way we are taught that science ought to work. A new hypothesis is aired, and used to make a testable prediction. Once the test is carried out, we decide whether or not the hypothesis is a good one, worth elevating to the status of a theory, according to the outcome of the test. In this sense, sociobiology is established as a set of good working hypotheses, or a scientific theory, unless and until it has to be modified, or replaced by a better theory, when it fails some future test of this kind. Wilson himself prefers to regard sociobiology as a scientific discipline, like physics, say, and in that sense the whole of sociobiology is not a scientific theory, and not testable in the same sense. Instead, it *contains* many testable hypotheses and theories within itself. But just as the success of Newton's laws helps to give us confidence in the whole structure of physics, so the success of this particular prediction helps to give us confidence in the whole basis of sociobiology.

Since 1975, the sociobiology approach has also led to a better understanding of reproductive strategies among animals – how mates are chosen and why some species are monogamous while others are not – and to the investigation of different strategies for care of the young, such as the difference between birds and frogs mentioned earlier. Why, for instance, is there a roughly equal balance between males and females in our own species? And a close cousin of games theory, called optimization theory, can, as Krebs explains, correctly predict '*exactly how much time* an individual will spend searching for food or mates in a particular place' (italics in original). Finally, for our present purposes, sociobiology has opened up another fruitful line of investigation, asking how much it benefits an individual to transfer information to others, either through its displays or through its vocal calls or songs. Giving away information is likely to be a bad strategy. In our example of the Hawks *v.* Doves scenario, for example, for two Doves staring each other down the worst thing to do would be to hint that you were about to cut and run. The new view of animal behaviour suggests that many signals are designed (that is, have evolved by natural selection) to provide *mis*information, to persuade, like advertisements. And since a deliberate liar can be unconvincing, the best way to mislead others may well involve the individual making the display believing in it: the Dove about to cut and run really does not know, until the last instant, that this is what is going to happen.

Important new insights into evolution at work are to be gained from the application of sociobiology, including the concept of the selfish gene and the use of games theory, to the world of plants. But we have more than enough here to chew on already, and anyone interested in following up all the new avenues of research can seek out Wilson's books and others cited in the bibliography. Here, we want to concentrate on animals, particularly mammals and specifically human beings. Four key features of human behaviour need to be explained by any good theory of evolution. The evolution of cooperation, or altruism, was for decades the most puzzling problem of all, and (as we shall see in the next chapter) the explanation of altruism was the success that laid the foundation stone of sociobiology. Apart from finding enough to eat, the fundamental requirement for all life, the three key sociological issues for people are: the problem of sex (how to find a mate, and what to do once you have found one); the need not just to mate and dump the offspring to fend for themselves (like many frogs), but to care for the and nurture the young (a problem particularly acute for human beings, with their long childhood); and the thorny problem of just how much innate aggression we really do carry 'in our genes'. Those issues, and the sociobiology that underpins the human variations on the three themes, form the basis of subsequent chapters. But before we get to the meat of our message, there is one lesser issue which has aroused great passions from time to time during the past hundred years or so, and which provides a nice case study of what sociobiology is and is not about. This is the question of intelligence.

Intelligence and evolution

You might think, to read some of the writings of the biological left, that sociobiology fails as a scientific discipline simply because it cannot account for the present variability of human intelligence (as revealed by standard tests) in purely genetic terms; that it has, in a sense, failed the intelligence test.* Far from it. Intelligence is one of those concepts which, like time, is impossible to define satisfactorily. We all know what intelligence is, just as we all know what time is, but none of us can put the concept down properly in words. This is the root cause of all the misunderstanding about intelligence, which has not been helped by the unfortunate use of the term 'intelligence quotient', or IQ, to describe an ability of the human brain that is tested by IQ tests. What ability is tested in this way? Only the

* *Not in Our Genes*, by Steven Rose, Leon Kamin and R. C. Lewontin, is a good example of this (false) line of argument.

ability of individuals to do IQ tests, a skill which is certainly related to 'real' intelligence, but is very far from the whole story, as we shall see. But before plumbing those murky depths, one thing should be clear. Unless you believe that God (or some god-like visitor similar to the ones in the film *2001*) was responsible for making the changes in our DNA that distinguish us from pygmy chimpanzees, or Belding's ground squirrel, or an ant, then it is clear that human intelligence, the most important aspect of being human, has been produced by evolution through the process of natural selection. As we have seen, this depends on two things. First, characteristics are inherited by individual members of a species from the individual's parents through, as we now know, the replication and passing on of genes, bits of DNA. Secondly, there is variation among the individual members of a species. And thirdly, natural selection operates because the individuals that are better fitted to their environment will be more successful, and will pass on more copies of their genes than those individuals whose phenotypes are less well fitted to their environment.

This applies to every characteristic that has evolved, from the relative length of your arms and your legs, to the innate human ability for language, to the efficiency of the haemoglobin in your blood as a transporter of oxygen. Of course, there may be characteristics (such as different eye colours) which are neutral, conferring no particular evolutionary advantage or disadvantage, and not subject to much in the way of selection. (We are not saying that eye colour is selectively neutral, only that it might be.) But few would argue that intelligence falls into that category. We are more intelligent than our ancestors, and we owe our success as a species in very large measure to that intelligence. Clearly, intelligence has evolved. Therefore, equally clearly, intelligence is something that is inherited. Certainly up until the emergence of modern man, *Homo sapiens*, individuals with above-average intelligence *must* have produced children with above-average intelligence, and they must have been more successful in raising those children, so that the genes 'for' intelligence spread through the population in succeeding generations. At the same time, less intelligent individuals must have been producing less intelligent children, and those children must have been less successful, in each generation, than their more intelligent rivals. If this had not been so, intelligence in the form that we possess it today would not have evolved. The selection pressure may not have been strong, and the difference involved may have been small (after all, they took millions of years to turn *H. habilis* into *H. sapiens*). But undoubtedly they did exist. Whether they still exist, and whether they operate in the same way today, is perhaps open to doubt. The human species now controls the environment to such an extent that characteristics that would cause individuals to be at a disadvantage even 50,000 years ago, let alone 5 million years ago, are not 'selected against' in the old way. In many ways, it is no longer a question of fitting ourselves to the environment, but rather one of fitting the environment to us. But the line of argument we have just outlined at least suggests very strongly that human intelligence is an inherited characteristic, and one which shows a degree of variability within the human population. Even so, although the semantics may be confusing, that is not at all what the often heated 'IQ debate' has been about.

The intelligence test

Between 1924 and 1974 some 7500 US citizens were sterilized by doctors acting on government orders, without the patients being told what was happening to them. The scandal came to light only in 1980; it was based on a misconception of the nature of intelligence and heredity that was unfounded in the 1920s, and had long been proved to be unfounded by the time this government-approved 'breeding programme' was halted. In every case, the victims were people regarded as mentally deficient, and many were themselves the offspring of parents with low intelligence. The pseudo-scientific basis for the sterilization programme can be summed up in the words of one Supreme Court judge in 1927: 'three generations of imbecility is plenty'.

You would think, from that, that there was a proven basis for the assumption that the overwhelmingly dominant factor in determining the intelligence of an individual is the intelligence of that person's parents. And yet, ironically, this attitude was fostered by the misuse of a system for measuring intelligence that originated, in France, from evidence for the exact opposite: that the important factor in determining the intelligence of an adult human being is the way that person has been raised. Nurture, not nature, determines the extent to which the intellectual potential inherent in all healthy members of *H. sapiens* is achieved.

Ideas about intelligence have tended to reflect the times. In the nineteenth century, the common belief (in Europe and the USA, at least) was that the white races were superior, and destined to rule the world, while, closer to home, it went without saying that men were more intelligent that women. The nineteenth-century French anthropologist and physician Paul Broca was a great one for measuring the size of any brain he could lay his hands on, and even he usually managed to delude himself into believing that the measurements supported his preconceptions. The average weight of a human brain is about 1450 g, so Broca was pleased to find that the famous Russian novelist Turgenev had a brain weighing more than 2000 g, while the zoologist Cuvier had a brain weighing 1830 g. The discovery that the great mathematician Gauss had a brain weight of only 1492 g was a little disappointing, and Broca could never know, of course, that his own brain weighed a mere 1424 g. But the way in which expectation moulds our view of the world is best shown by his study of the brains of several German professors, late of Göttingen University. Finding that some of them were relative lightweights in brain size, he wrote 'the title of Professor does not necessarily guarantee genius'. He could, of course, have concluded that being a genius does not depend on having a large brain!

Broca's attitude to women similarly reflected contemporary attitudes of his era. He noted that women have, on average, smaller brains than men (the different is about 180 g), and commented, 'we must not forget that women are on average less intelligent than men,' not totally failing to point out that women are, on average, smaller than men, but continuing, 'the relatively small size of the female brain partly reflects her smaller physique and partially her intellectual inferiority'. In fact, as Broca could easily have calculated, smaller physique alone is more than enough to account for women's smaller brain size. But, as we now know, brain size is not simply correlated with intelligence. It is the quality of the brain, not just the quantity, that counts.

Similar prejudices shine through in other studies of the supposed intelligence of underprivileged groups, such as blacks, in nineteenth-century studies. There is no point in elaborating on them here, for the attitude to women as expressed by Broca is entirely representative of the times. So it came as a shaft of scientific light among the gloom of all these preconceptions when the Frenchman Alfred Binet tackled the problem of 'backward' schoolchildren in the early 1900s.

Binet was working with children who had learning problems. Such children could not benefit from the standard teaching programme used in the schools of Paris, and the initial problem facing Binet was to devise a test which would weed out these misfits and ensure that the rest of the children proceeded smoothly with their education. But why were some children backward? Binet argued that the problem was not that they were innately stupid, but that, for one reason or another, their intelligence had not been brought out, and had failed to develop. Once Binet's early intelligence test had identified a child with 'low intelligence', the obvious thing to do was to provide that child with a special educational programme in order to exercise his or her intellectual skills and enable it to catch up with other children of the same age. Just as a physically underdeveloped child could be assigned a programme of athletics and weight-training exercises to build up the body, so, Binet reasoned, a mentally underdeveloped child could be given equivalent mental training.

Binet's test started out from the assumption that older children should be able to carry out mental tasks that younger ones cannot, and he decided on the 'normal' level of achievement for each age group by discussing the children's abilities with their teachers. He developed a simple test that produced scores, when those children were tested, in close agreement with the teachers' estimates of the relative intelligence of the children in their charge, and then used the test to identify children who were two or more years 'behind' their peers and needed special tuition. And Binet's ideas were supported by evidence that came from Belgium when his test was first used there. Children in a Belgian school were found to have much higher 'mental ages' than the children in Paris: exactly what should have been expected, said Binet, since the Belgian children were showing the benefits of attending a private school with small classes, as compared with the Paris children, who attended larger schools with much bigger classes, and enjoyed less individual attention from teachers.

The Binet test, first published in 1905, should have ushered in a new era of educational enlightenment. Instead, translated and adapted for use in the English-speaking world, it became an instrument used to support the entrenched nineteenth-century preconceptions about intelligence.

Binet himself died young, in 1911, and was not around to protest at the transformations wreaked on his test, and its misapplications. In the USA the test was translated and adapted by Lewis Terman, of Stanford University, who introduced what became known as the Stanford–Binet test of 'IQ' in 1916. This was essentially the test that was to be used to determine the fate of hundreds of thousands of people in the USA over the next half-century, together with a similar test devised by US psychologist Robert Yerkes. The tests perpetrated such inanities that it is hard to see, looking back from the 1980s, how they were ever taken seriously. Polish immigrants, with no knowledge of American language or customs, were classified as having low IQ partly because they were unable to name famous baseball stars (we kid you not); illiterates, who could not take a written test, were assessed by, among other things, showing them a picture of a tennis court with the net missing, and asking them to complete the picture. Apart from the fact that many had never held a pencil before, many had never seen a tennis court in their lives. Result: a score of zero, and another imbecile is added to the list. Thanks largely to these tests, it was officially decided that Poles and Italians were genetically inferior to other would-be immigrants, too stupid to be admitted to the USA. And at a time when many were desperate to escape from the rising tide of Fascism and Nazism in Europe, in the 1930s the doors of America were closed to them. Quite literally, innocent people died because of the Stanford–Binet and Yerkes tests.

In England, the Binet test was enthusiastically taken up – and misused – by Cyril Burt, to such effect that late in life he received a knighthood for his contributions to remodelling the

British educational system. Burt was largely responsible for the iniquitous 'eleven-plus' system (which we both benefited from, but was no less iniquitous for that) whereby children were tested, with a kind of Binet test, at about age 11. No account was taken of the children's background. As a result of that one series of tests, they were labelled as having a certain intelligence, and those with a score higher than an arbitrary cut-off point were selected to enjoy the advantages of a 'good' education leading towards public examinations and, hopefully, to university. One of his aims, to be sure, was to bring through the system into higher education children who were intellectually able but came from a class background which would have deprived them of educational opportunities. That seems laudable enough; but the rest were dumped into second-class schools, where they received fewer opportunities and, hardly surprising, achieved less. This was the exact opposite of Binet's approach. He would have said, it now seems correctly, that the children who 'failed' such a test should have been those to receive special attention, and that as a result they would then very probably have achieved, in due course, as much as those who had 'passed' the test.

Burt's case is a sad one of self-delusion, based on preconceptions so extreme that they led him, researchers discovered after his death, to invent data and publish fictitious scientific papers under false names (sometimes in journals he edited). The papers were entirely made up out of his own head, to back up his ideas. He did not intend to mislead, of course, but simply could not be bothered to do the research, because he 'knew' what answers it would give. In the 1950s nobody checked up on him, because 'everyone knew' that IQ was an absolute, a fixed, unchangeable number carried by individuals throughout life. Unfortunately that is still part of popular mythology, and education in Britain is still suffering the consequences. As recently as 1973 psychologist Hans Eysenck could be found making the claim – described by Nobel Laureate Peter Medawar as 'the silliest remark I have ever heard' – that 'the whole course of development of a child's intellectual capabilities is largely laid down genetically'.*

* Quotes from *Nature*, 19 July 1984, Volume 310, p. 255.

One of us is a teacher, and has seen first-hand how the myth typified by Eysenck's remark can still warp lives. Time and again children in schools in the 1990s are labelled as virtually unteachable, and in overcrowded classrooms are largely ignored, left to amuse themselves as best they can while harassed teachers concentrate on teaching the majority as much as they can. Yet, in our experience, every time such 'unteachable' children have been singled out for attention, they have blossomed, soon catching up to the point where there is some hope of including them in the routine classwork, and often going on, having discovered unsuspected abilities, to become one of the class leaders. With classes of thirty or more, there is a limit to what can be achieved without neglecting the other children. But in an ideal world where children were taught in groups of four or five, each group receiving the full attention of a teacher, then, our own experience suggests, Binet's expectations would be amply fulfilled.

Many genuine studies, including those of adopted children, children in orphanages, and twins raised apart and together all point to the same conclusion: given the right environment and mental stimulation from infancy, any child will score well on a Binet test by the age of 11, and will grow up to be an intelligent adult. It is an indictment of society, not genes, that so many children today fail to achieve their potential. And this brings us right back to the heart of what sociobiology is all about.

Sociobiology, heredity and intelligence

Human sociobiology is not about subtle differences between one human being and another. Whether Linford Christie can run a hundred metres faster than Carl Lewis (or whether either can run it faster than the authors of this book) is not a matter for sociobiological debate. What is interesting is that both these great athletes, and both the authors of this book, and all other members of the species *H. sapiens*, do their running on two legs, not on all fours. In other words, the comparisons that should be made are those between *H. sapiens* and other species. Sociobiology may help us to understand why the chimpanzee lifestyle is successful under certain conditions, why frogs in general do well in the kind of environment frogs inhabit, and why people are so successful compared with other species. From this perspective, we see immediately that intelligence is a huge advantage for *H. sapiens*, and we see that it is an inherited characteristic that has evolved markedly over the past 5 million years or so. We see that *all* normal human beings inherit the capacity to be quite astonishingly intelligent compared with all other species on Earth, even the chimpanzee. Even a human being from a deprived environment is much, much smarter than the brightest chimpanzee. *That* is what sociobiology is all about.

Of course, there are variations in intelligence, just as some people are bigger than others and some run faster than others. The occasional genius like Albert Einstein is offset in the statistics by the occasional moron. But, by and large, every individual human being has the innate capacity to be as intelligent as any other, just as we are all born with the innate capacity to become fluent linguists. Very many people grow up with stunted linguistic skills because they are brought up in almost inarticulate surroundings; very many people (often the same people) grow up with stunted mental skills, because those surroundings are also intellectually stunted. But the lesson we learn from sociobiology is that if society cared to provide the proper opportunities for all its citizens, they would all develop to what is nowadays regarded as a high standard of intelligence.

It is supremely ironic that the abuses of the Stanford-Binet and Yerkes tests, and the work of Cyril Burt, are high on the list of reasons why Marxists, whose creed is one of equality, have attacked the whole concept of human sociobiology. Genetics and sociobiology tell us that human beings are indeed created very nearly equal, just as the Declaration of Independence tells us, sharing the vast majority of their DNA, and each with the potential to achieve great things – as athletes, musicians, linguists, mathematicians or whatever – depending on whatever very slight natural preferences they may have to go in one direction or another, and, crucially, on the opportunities they are given to develop those innate abilities. And this unitary view of humanity is the underlying theme of the whole of sociobiology, and of the rest of this book. When we look at the relationship between parents and children, for example, we are not concerned with the question of whether Mrs Jones is a 'better' mother than Mrs Brown; we are interested in finding out why any human mother should be 'willing' to devote so much time and effort to raising a helpless infant. How could such a system have evolved? When we look at sex, we are not really bothered about whether Mr Jones is a more faithful husband than Mr Brown; we want to know why human beings pair up in marriage, why there should be two sexes, and why the numbers of men and women should be equal. When we look at aggression, we are not going to puzzle too much over their question of whether one individual or another is violent; instead, we look at the overall patterns of human aggressive behaviour. And when, later in this chapter, we puzzle over the nature of altruism, we are not going to worry about whether or not one person is kinder than another. Of course there are variations within the whole population, just as there are with most inherited characteristics. The really interesting thing is that people have worked together so effectively, over the generations, that we have come to dominate the planet in less than 10,000 years since the invention of civilization. How can this be squared with the notion of old-fashioned Darwinism, with individuals struggling for survival and for scarce resources in fierce competition with other individual members of the same species? And how does altruism fit in with the 'selfish genery' of Richard Dawkins? The simplistic interpretation of those ideas leads you to expect that every man will be out for himself, every woman likewise, and the devil take the hindmost. The surprising thing is that the vast majority of people are so nice to each other. Sociobiology can tell us why.

Fitness, in the Darwinian sense, is measured by the success of an individual in reproducing, in passing on copies of its genes to the next generation. Strictly speaking, the only way to compare the relative evolutionary success of two individual members of the same species is to wait until both of them have died, and then to count up the number of offspring they have left behind to reproduce in their turn. The more surviving offspring, the more 'fit' the individual was.

For species like our own, with two sexes, there obviously has to be some minimal degree of cooperation between two individuals – a man and a woman – in order to reproduce at all. This vital aspect of being human is the first topic we shall look at in detail. But, apart from the necessity of finding a breeding partner, naive Darwinism might lead you to expect individuals to be always in conflict, often violent conflict, with other members of their own species. Although a lion cannot really be said to be in competition with, say, a swift, since they occupy quite different ecological niches, a lion must be in very real competition with other lions since they live in the same parts of the world, eat the same kind of food, and even need to find the same kind of mates. Close evolutionary relatives ought to be most competitive, on this simple picture, because their needs are very similar and they 'ought' therefore to be fighting over the same available resources.

Against this picture of naive Darwinism, the fact that people sometimes act aggressively towards one another, and that human history is dotted with conflicts, is no surprise at all. But it is absolutely astonishing that we are able to cooperate with one another, that we band together in tribes, cities, nations and other sociobiological groups to work, by and large, for the common good. We are, indeed, civilized. Although many scientists and popularizers have made much of the aggressive side of human nature, it does remain the exception to the general rule, and this is why it is newsworthy (how often do you hear on the news about an old lady who walked down the street without being mugged?), and it is fascinating to us largely because it is not 'normal' everyday behaviour. Biologists will tell you that humans do not, in fact, rank very high on the scale of fighting, aggressive mammals. There are many hunters more vicious than ourselves, and several species in which conflict between individuals is much more common and bloody. Even deer, in spite of our cosy image of cuddly Bambi, go in for spectacular combat at certain times of the year. No, there is no getting away from it: people are quite nice, really. But why?

In his *New Scientist* article in 1976, Edward Wilson singled out altruism as 'a central theoretical problem in sociobiology'. Altruism is defined very carefully by biologists. An altruistic act is one that gives a benefit to one individual at a cost to another, and it is measured in reproductive terms, so that an altruistic individual is one that reduces its own reproductive success in order to enhance the reproductive success of another individual or individuals. The extreme examples of this are found in social insects such as bees, where there are whole castes of sterile workers who never reproduce at all, but spend their lives ensuring that the queen breeds successfully. This behaviour is now very well explained in terms of genes and sociobiology, as we shall see. But in more general terms, altruism can extend right up to the level of just plain being nice to people, loving thy neighbour, in fact. Individuals have evolved through billions of years of natural selection, and only characteristics that are advantageous in terms of evolutionary fitness (or, at the very least, confer no evolutionary burden) will survive the winnowing process. We share 99 per cent of our DNA with other animals, and our 'niceness' must have been selected for; by some measures, we are the most successful species on Earth, and there is every reason to think that our niceness is a major contributor to that success. If we did not cooperate, we would reduce our own numbers so effectively that other species would have more opportunities. It looks as if the natural aggression and competition that the simplest version of Darwinism predicts has been suppressed by some mechanism which acts for the benefit of the species, or of a large group of individuals. This is a superficially appealing idea which still appears from time to time in some popular accounts of evolution. But group selection was knocked on the head, as a scientific theory, in the early 1960s.

Ironically, but in the best traditions of the scientific method, group selection was clearly seen to be inadequate as a description of evolution at work once one painstaking scientist had carried out a thorough analysis and published a comprehensive review supporting the group-selection hypothesis. Once the hypothesis was laid out logically, other researchers were able to pinpoint the flaws in the argument which led to a revival of the strict Darwinian view that selection operates on individuals, and that was a major influence in the rise of sociobiology. The difference from Darwin's day, however, is that we can now see the importance of genes – which were unknown to Darwin – to the evolutionary story. Individual phenotypes will behave in such a way that they maximize the chance of copies of the genes that they carry being passed on to the next generation, whether those genes come from their own bodies or from those of another member of the same species.

Richard Dawkins's image of the 'extended phenotype' is the modern version, gene selection, which sees members of a group cooperating because they share many genes in common. But the perspective is exactly the opposite of the original group selectionists, starting out now from the smallest reproductive units and working upwards, instead of starting out from the species and breaking it down into individuals.

The group-selection fallacy

It is worth taking a quick look at the ideas of group selection, since an understanding of the flaws in the argument leads almost inevitably to an understanding of how natural selection really does work. The definitive statement of group-selection ideas was made by the Scot V. C. Wynne-Edwards, in his 1962 book *Animal Dispersion in Relation to Social Behaviour*. Wynne-Edwards argued that animals have a natural tendency to avoid overexploiting their habitats. In particular, he said, individuals have some means of recognizing how much food there is to go round, and then, for the good of the species as a whole, they will hold back their reproduction, or even give it up altogether, to avoid overpopulation. In this picture, a great deal of social behaviour was thought to stem from the need for individuals to be kept informed of the population density. The gathering of flocks of birds or herds of deer, for example, was supposed to allow individuals to assess the numbers and adjust their own reproductive activity accordingly.

The counter to this line of argument in terms of individual selection is very straightforward. Just as in the Hawks *v.* Doves scenario, imagine a single mutant individual occurring in a large group of, say, birds. Every other member of the group might be a good, well-behaved group selectionist, busy counting up the size of the flock and deciding that there are too many individuals around this year, so it would be best not to lay any (or not many) eggs. The mutant, however, has no such qualms. It sets out to rear as many offspring as it can. Some of those offspring will carry the same inherited tendency to maximize their own reproduction, regardless of the good of the group, and the genetic package that makes for this behaviour will spread rapidly in succeeding generations. The mathematics of

Maynard Smith's games theory confirms what ought to be intuitively obvious, that the mutant behaviour soon becomes the dominant pattern: the ESS in this case, as in almost every other case, is to rear as many offspring as possible at all times.

This does not mean, for a bird, laying as many eggs as possible at all times. Studies of the English great tit, for example, have shown that although the typical clutch of eggs numbers 9 or 10, some birds lay fewer and some as many as 12 or 13 eggs. But under normal environmental conditions (average weather, and so on) the largest number of nest-mates surviving to be fledged from the nest is always from the nests with 9 or 10 eggs to start with. If there are too many offspring for the parents to feed adequately, the weakest go to the wall and fail to reach maturity.

Problems also arise with group-selection ideas when we look at the evolutionary process. Evolution by natural selection depends on the difference in reproduction between the successful and the less successful. According to the group-selection hypothesis, that means that more successful *groups* have to be selected for, while less successful groups die out. This is difficult enough to arrange in any mathematical model (let alone a plausible one); it becomes virtually impossible to arrange if we consider the possibility that when a group is facing extinction, for whatever reason, then natural selection will strongly favour any individuals who move away from the threat and continue the existence of the species elsewhere.

In the 1990s, the balance of evidence clearly supports the original idea of individual selection rather than group selection. Robert Trivers has summed up four key points against Wynne-Edwards's ideas;* they can be paraphrased like this:

1 There are no known patterns of behaviour that cannot be explained more easily in terms of natural selection of individuals, so group selection is unnecessary.
2 Outside influences, such as food shortages or predators, control the numbers of a species present in any region without the kind of 'group altruism' required by Wynne-

* Robert Trivers, *Social Evolution*, p. 81.

Edwards, and without the kind of group extinctions required if selection were acting at the level of the group.

3 Since the early 1960s, many studies of the breeding success of individuals – their evolutionary fitness – have been carried out, and they *always* show that animals adopt the strategy of reproducing as rapidly as circumstances permit.

4 Although mathematical games can be set up where group selection operates, it is always a weak effect which is at the mercy of the arrival on the scene of a mutant individual acting in its own self-interest. Group selection *never* provides an ESS.

In the world today, and throughout the vast bulk of evolutionary history, all new traits must appear in rare members of a species, says Trivers, and will increase in frequency among the members of the population only if they increase the survival prospects and reproductive success of the individuals carrying the new traits.

Darwin up to date

Even what seems to be altruism, operating at the level of individuals – or phenotypes – is really selfishness operating at the level of genes. The important feature of reproduction, in evolutionary terms, is not that a human being gives birth to another human being, or that a wasp hatches from a wasp egg. What matters is that DNA is being copied, and a gene is the basic unit in that process. The argument is most familiar today through the widespread success of Richard Dawkin's book *The Selfish Gene.* It was actually put on a secure mathematical footing in the early 1960s (ironically, at about the same time, but with far less immediate attention, that Wynne-Edwards's ideas on group selection were publicized), by William D. Hamilton, and the basic concepts had already been aired by the mathematician Ronald Aylmer Fisher and by J. B. S. Haldane. Selfish genery is Darwinism at its simplest, and, in its latest form, Darwinism at its most up to date.

'Traditional' Darwinism sees evolution acting upon individual phenotypes, which are the units of selection. It is in this sense that altruism is a puzzle, because it is not immediately obvious why an individual should help another individual to reproduce at the cost of reducing his or her own chances of reproduction. The blindingly simple solution to this puzzle – a very real puzzle before the genetic mechanism of heredity was understood – comes from looking at close relations and at their common genetic inheritance. The closest relations are parents and their offspring, and full siblings (brothers and sisters who share the same parents), so this is the best place to start.

In a sense, even parental behaviour is a form of altruism, especially in a species like our own, where infants are so helpless for so long. Of course, it is obvious why care of helpless infants should have evolved: if we don't look after our own children, we have very little prospect of passing on our genes to succeeding generations. Especially before the invention of civilization, genes that lead people to be good parents will have survived, while those that make for individuals that neglect babies will not have. And it is by extending this idea that we can arrive at an understanding of altruism in general.

We might as well think in human terms here, although what we have to say applies to any species that reproduces in the way we do. Each new human being has a unique genotype (excluding the special cases of identical twins, triplets, and so on) which is made up of a combination of genes inherited from each parent. In most cases the mother and father are not close relations, so we can regard each of them as having an independent set of the possible genes (strictly speaking, alleles) available from the human gene pool. Half their child's genes come from each parent, so geneticists say that there is a relatedness of $1/2$ between a child and either of its parents. The same simple rule applies to all children and all parents: the relatedness between father and daughter, mother and son, mother and daughter and father and son is always $1/2$. What about the children's relatedness to each other?

All children of the same mother inherit half her genes, but they do not all inherit the same genes from her. It is possible that two of her children may inherit copies of almost exactly the same half-set of genes, and therefore have very similar genotypes; equally, it is possible that one may inherit half her genes and the other may inherit the other half, reducing the relatedness. On average, though, there is a 50–50 chance that one of the genes copied and passed on to one child will also be passed on to its sibling. So, on average, half the genes each child inherits from the mother will be the same as those in any other child that has the same mother. And, since half the total genotype is provided by the mother, then on average siblings that share the same mother have a relatedness of $\frac{1}{4} = (\frac{1}{2} \times \frac{1}{2})$. Exactly the same reasoning applies to siblings who share the same father. So for full siblings, which share *both* parents, we simply add up the two contributions to find their relatedness is, on average, $\frac{1}{2} = (\frac{1}{4} \times \frac{1}{4})$, the same as the relatedness of parent to offspring.

It is very simple to extend this argument to other relations. The next closest kin, after siblings and parents, are full cousins. You share half your genes with your mother, who, on average shared half her genes with her sister, who, in turn, shared half her genes with her daughter. So the probability that you and your cousin share any one gene is $\frac{1}{2} \times \frac{1}{2} \times \frac{1}{2}$, or $\frac{1}{8}$. This kind of calculation immediately puts a new complexion on Darwinian evolution.

If an individual helps a close relation, such as a sibling or a cousin, to reproduce, then that altruistic act is actually helping many copies of genes that are also present in the altruistic individual's body to reproduce. Sometimes the genes that are caused to spread in this way will include copies of the genes that made the altruistic individual act in this helpful way. And that, plus some detailed mathematical calculations by the likes of Hamilton, Trivers and Maynard Smith, is all you need to understand altruism in Darwinian terms. A gene or set of genes that causes individuals to behave in what we think of as an altruistic way can spread through the population because its presence in some bodies (phenotypes) will lead to action which allows copies of that same gene to reproduce in other bodies. This is all best summed up by a remark that Haldane is supposed to have made in a pub in the 1950s. Discussing the puzzle of altruism over a pint or two with some colleagues, he was asked if he would lay down his life for his brother. Haldane throught for a while, and then replied, according to legend, 'Not for one brother. But I would for two brothers, or eight cousins.' The point is that, on average, behaviour that ensures the reproductive success of two of your full siblings, or eight of your cousins, ensures the survival of copies of every gene that you carry, simply by adding up the relatedness. It is as good, Haldane realized, as if you lived to reproduce yourself.

Of course, Haldane's remark should not be taking too literally. People, let alone other animals, do not sit around calculating degrees of genetic relatedness before deciding whether to jump into a river to save a drowning child. But the fact that people are more likely to risk their own lives to save a close relation is so firmly established that, for example, the award of the Carnegie Medal is specifically excluded for such acts. If you want to win this bravery award, you have to save the life of someone outside your immediate family.

Once we get down to specifics like this, many people find it hard to accept that human behaviour really does depend largely on this kind of genetic inheritance. So before we go on to look in more detail at altruism in people, it may help to see how beautifully these ideas stand up to the test in other species.

The birds and the bees

If at first you find these ideas hard to swallow, you are in good company. When Hamilton, graduated from Cambridge University in 1960, and began working for his PhD at the University of London, the group-selection school of thought was very much the dominant one in the evolutionary debate. Hamilton found it hard to reconcile this with the precise mathematical treatment pioneered by Fisher to explain how evolution worked by selection at the individual level, and he tried to bring Fisher's ideas up to date, and make them more complete, in his PhD thesis. Unfortunately, the powers that be decided that this work was not up to the standard they required, and Hamilton was told that he could not receive a PhD for it. Nevertheless, his work was published in 1964, in a paper that Trivers has described as 'the most important advance in evolutionary theory since the work of Charles Darwin and Gregor Mendel'.* Even without his PhD,

* Robert Trivers, *Social Evolution*, p. 47.

usually the essential meal ticket for a university scientist, Hamilton became a lecturer at University College, London, in 1964, and at last received his doctorate in 1968. Indeed, recognition came much quicker to him than to Mendel, who never lived to see the importance of his own work recognized.

Why is Hamilton's work now seen as so important? We do not want to become too deeply embroiled in the mathematics here; you can find the details, if you wish, in student texts such as David Barash's *Sociobiology and Behavior*. But insect communities, such as those of the bees, provide such a striking example of the success of modern ideas about gene selection that they are worth examining in a little detail.

Darwin himself puzzled over the behaviour of insects. In the first edition of the *Origin*, he commented on:

> One special difficulty, which at first appeared to me insuperable, and actually fatal to the whole theory. I allude to the neuters or sterile females in insect-communities; for these neuters often differ widely in instinct and structure from both the males and fertile females, and yet, from being sterile, they cannot propagate their kind.*

The problem is simple. We have what seems to be the ultimate in altruism, workers who work hard and diligently to ensure the successful reproduction of the queen, who produces thousands of offspring, while the workers leave no offspring behind at all. How can evolution by natural selection explain such a phenomenon? How can it possibly be 'fit' to fail to reproduce? Workers leave no offspring, so where do the workers in the next generation come from? Or, in modern terms, where do the worker genes in the next generation come from? Stated like that, we are already on the trail of the answer. The genes come, of course, from the queen, and from the male she mates with. But social insects are not among the category of creatures that reproduce in essentially the same way that people do. Males in these species (which include bees, ants and wasps) have only one set of genes, while the queen has two. And this makes all the difference.

* See pp 203–4 of the Modern Library edition. Darwin follows up with his own solution to the puzzle, which is very similar to modern kin-selection theory.

When a queen bee produces sons, she does so without any genetic contribution from a male. Her eggs are formed by cell division so that each contains one set of genes. If the egg develops without fertilization (parthenogenetically), it becomes a male. But when she produces daughters, the eggs are fertilized in the usual way, and end up with two sets of genes. Because the male – the father – has only one set of genes to copy and pass on, however, this makes the daughters much more closely related than are human siblings. The relatedness of mother and daughter is just as before, $\frac{1}{2}$. But the father now gives a copy of *all* his genes to the daughter, so the relatedness between father and daughter is 1: every gene in the father is copied in the daughter's genotype. All the daughters have the same father, because in such species the queen stores up the sperm provided by her partner in one mating and shares it out among her eggs as they are laid. So half the genes in all her daughters are identical, because they come from the father; and half the other half are identical, exactly as with human mothers and daughters. So the *total* relatedness between sisters in the hive is $\frac{3}{4} = (\frac{1}{2} + \frac{1}{4})$. The relatedness between sister and brother is only $\frac{1}{4}$, since the brother has no genes from the father and is in effect a half-sibling, while the relatedness between brothers is $\frac{1}{2}$. But it is the sisters that matter here.

As Hamilton pointed out, and calculated explicitly in his landmark contribution of 1964, the relatedness of $\frac{3}{4}$ between sisters in the hive is more than the relatedness that would exist between mother and daughter if the workers were to reproduce in the usual way. If a female worker were to reproduce in the way that we do, she would pass on half her genes in every new individual born. But if, instead, she helps the queen to reproduce, then on average she ensures the reproduction of three-quarters of her own genes every time a new female is born. In terms of selfish genes, it is more advantageous to give up reproduction for herself and act 'altruistically' towards her queen and sisters.

Extension of this argument explains, in detail we shall not go into here, why there are more females than males in the hive, why the males do relatively less work than the females (with a relatedness of only $\frac{1}{4}$, it does not pay them to help their sisters) and many other features of the life of social natural selection really could be applied universally became, in the hands of Hamilton, a triumphant example of evolution and selection at work at the level of genes.

We have not, of course, gone into the fascinating question of how such a genetic system could have become established in the first place. As we said at the start, this book is concerned with explaining animal behaviour, especially human behaviour, today, in terms of our genetic inheritance. The origin of species is outside our scope. What matters here is that modern Darwinism explains the continued survival of the way of life of the social insects today, a way of life that has evolved independently in twelve varieties of present-day life on Earth, including the termites, which is some measure of its success.

Bees and ants, though, are far removed from human beings, and from human behaviour, as this modern explanation of an old problem itself shows. But some of the closest parallels with human behaviour come from species that are not the ones that most of us would immediately think of as models for ourselves – birds.

The search for natural 'models' on which to base human sociobiology leads us up some strange paths. Social insects very definitely are not organized like human society, yet their peculiar genetics provides the basis for a thorough test of sociobiology ideas, tests which sociobiology passes with flying colours. To study one of the most characteristically human features of our social interactions, we have to turn not to our closest genetic relations, the chimp and the gorilla, but to the birds. The feature of bird behaviour that is shared with the behaviour of people, but with very few other species, is that the male usually makes at least some contribution to the care of his offspring, and often plays a major role in rearing the young. This is certainly not the case in chimpanzee society, for example. Birds, like people, have what the biologists call 'a high male parental investment', and this carries with it a host of associated traits that sometimes echo our own behaviour and sometimes provide new insights into the business of being human. This is one of the principal reasons why evolutionary biologists spend so much time studying birds (the other reason is simply that there are many birds around and they are relatively easy to study). So what can the study of birds tell us about the evolution of altruism, and the reasons why it is so often, in one form or another, an ESS today?

One example will show the importance of this kind of work. Ornithologists were puzzled for many years by the observation that in many species of bird it is quite common for a young (but sexually mature) adult to make no attempt at breeding itself, but to devote a great deal of energy to helping an older pair to rear their own offspring, chiefly by carrying food to the nest. This helping at the nest has now been seen in more than 140 species; it is far from an odd freak of nature, and must represent a successful strategy, at least under certain circumstances. The puzzle is, of course, very similar to the puzzle of worker bees, since it seems that the helper is missing out on the opportunity to send copies of its own genes into the next generation, while ensuring the spread of rival genes. But there are subtle differences from the insect case, as well as similarities.

First, the similarities. As you might by now expect, the helpers are usually close relations of the individuals they are helping. Very often, helpers are previous offspring of the parents, and therefore siblings to the nestlings, that they help. So there is an immediate genetic benefit, which must be offset against any reproductive cost to the helpers, in the copies of their own genes that they are aiding in the bodies of their siblings. With birds laying several eggs in one nest, it very quickly becomes possible for a variation on Haldane's quip to become a real proposition: helping three or four siblings to survive might very well outweigh the cost of not breeding yourself.

But once this pattern of behaviour becomes common, as genes for helping spread through the gene pool of the species, it can lead to more widespread effects. The pattern of behaviour that makes for helping at the nest of your siblings is almost exactly the same as the pattern of behaviour that makes for helping at the nest of any members of your species. As helping becomes a common activity, some individual helpers, either through confusion, inability to recognize their kin, or even as the result of a very slight mutation, will help more widely. In one study of Florida scrub jays, for example,* 199 helpers were observed during an eight-year period. Of the 199, 118 were helping both their parents to raise their full siblings, 49 were helping one parent and a 'step-parent' to raise half-siblings, and the other 32 were helping various combinations of siblings, grandparents and unrelated birds. This begins to show how a characteristic that starts out on the basis of kin selection can spread through a population to become, for want of a better term, general 'niceness'. And of course, such a tendency then begins to act 'for the good of the species', although it has its origins firmly in the selfishness of genes.

The most important feature of all these studies is that birds with helpers do indeed rear more young from the nest than birds without helpers. But kin selection is not the whole story. Unlike worker bees, the helpers at the nest are not sterile and could reproduce. Many of them do indeed reproduce, but only after 'learning their trade' as helpers. Some studies show that young birds who find a mate and breed as soon as they are sexually mature do not produce very many fledged offspring. Hardly, surprisingly, parents with some experience of the task do a better job of rearing their young. In species where helpers do their learning at the nests of their close relations, there is no waste of effort in trying to rear offspring that fail to reach maturity, but the skills that are learned will prove just as valuable in later years when the bird finds a mate and sets up a nest of its own. Indeed, observations of groups of these birds over several years show that birds that have been helpers in previous years and then become nest-holders themselves do better in the reproductive stakes than young birds nesting for the first time without the benefit of previous experience.

And there is yet another parallel with human behaviour among these species. In some cases, including the Florida scrub jays already mentioned, successful helpers eventually inherit breeding space from the holders – often their own parents – when the older birds die. As we shall see, birds really do provide a useful model for some kinds of human behaviour. Altruism

* Discussed, among others, by David Barash in *Sociobiology and Behavior*, p. 89.

derived from kin selection really does have a genetic basis, while altruism of a slighty different kind may have played a big part in making us human in the first place.

The altruistic ape

Kin selection can explain the basis of many aspects of human altruistic behaviour. Our ancestors, in the not so distant evolutionary past, lived in small groups, a few dozen at most, in which everybody was somehow related to everybody else. In those circumstances, any tendency towards general altruism, provided it did not involve too much effort to risk on the part of the altruist, would tend to be a good thing for the genes carried by the altruist, since the survival of bodies carrying copies of those genes would be enhanced every time the altruist helped a relation. The key, of course, is the balance between cost and benefit, which has been analysed in great mathematical detail in recent years. In simple terms, we would expect people to be more willing to help close kin, and not to be willing to put the same effort (or take the same risks) to help more distant relations. This is certainly true: one of the enduring features of human society is nepotism in one form or another. And a situation rather more like that faced by our recent ancestors occurs even now in some parts of the world.

One example the sociobiologists quite rightly love to quote is that of the Aleuts, who live on the shores of the Bering Sea and go hunting for whales from open boats. Whaling crews consist of individuals who are close relations: brothers, fathers and sons, uncles and nephews. Whaling is a risky business, and the most successful crew will be the ones where individuals are willing to help each other even at some risk to themselves; everyone has to pull his weight, whatever the circumstances. An outsider will be less inclined to help a crew mate out of difficulties. Such examples are good circumstantial evidence in support of kin selection as a basis for altruism. But they are no more than the icing on the cake of the detailed studies of many species which show the predicted pattern of behaviour. And they do not tell the full story, since one of the key principal features of being human seems to have derived from another kind of altruism altogether, a seemingly cold-blooded calculation of how to maximize self-interest, which, ironically, is what goes towards making people so nice.

Wilson has described altruism as 'the mechanism by which DNA multiplies itself through a network of relatives'.* But Trivers has shown that this network can be extended much further, and DNA reproduction enhanced beyond the circle of relatives (except in the sense that all people are related to one another by common descent from some remote ancestor) by a process called reciprocal altruism. The essence of this is simplicity itself, familiar to all human beings, and can be summed up in a time-worn sentence: 'You scratch my back and I'll scratch yours.'

Reciprocal altruism provides the precise counterexample to aggression when we look at human society. Extreme aggression is fascinating because it is aberrant; reciprocal altruism is so 'obvious' to us that we take it for granted, like the little old lady who goes quietly about her daily routine without ever being mugged and appearing on TV. It requires a conscious shift of mental gears to understand why reciprocal altruism should be such an exciting topic in sociobiology. Our whole society is based upon reciprocal altruism. Economics depends on people being willing to accept promises from each other; without trust, our financial institutions would come crashing down. Of course there are cheats – in a society of mainly altruists, it may be that cheating provides a viable way of life (an ESS) for a few. We try to minimize cheating by having rules of behaviour, carefully worded contracts, all the machinery of justice, law and order, and police. But that is only a superstructure we build on an underlying framework of innate human nature. If most of us were *not* honest reciprocal altruists, the whole system would fall apart.

Opponents of sociobiology who fear that analysis of the genetic bases of behaviour might, for example, encourage judges to give lenient sentences to rapists because it is only 'human nature' have undoubtedly got hold of the wrong end of the stick. Our interpretation of sociobiology actually suggests

* Edward O. Wilson, *Sociobiology*, p. 58.

that very severe sentences should be handed out for such *un*natural (as far as the majority is concerned) behaviour, thus helping to minimize deviation from the pattern of behaviour that is most appropriate for human society.

Or consider our care of the sick, something more like the image that initially springs to mind when we hear the word 'altruism'. Naive Darwinism would suggest that the sick should be left to die. Why help them to recover and breed, thereby ensuring survival of their genes, when you could grab their resources for yourself and use them to provide for your own children? Tending for the sick is, however, an important feature of human behaviour, and has been since long before the rise of our modern civilization. We pay taxes, or subscribe to insurance schemes today, partly out of our self-interest, so that when *we* are ill we can receive the necessary medical attention. But most of us have very little prospect of needing major medical attention in our youth, while we are busy raising our children. By each contributing a relatively modest amount, we ensure that the few individuals who do need this kind of help get it. The cost to the altruist – the donor – is small, but the benefit to the recipient is high, perhaps literally a matter of life and death. This is a key feature of the insight into reciprocal altruism provided by Trivers in the early 1970s.

The most readable summary of the detailed application of these ideas to many species is to be found in his book *Social Evolution*. The relevance to human beings can be stated quite briefly, starting out from one of Trivers's examples: the case of a drowning man rescued by an unrelated bystander. This happens comparatively frequently in human society, sometimes at great risk to the rescuer, and occasionally with the would-be rescuer drowning as well. There is a range of human reaction to the sight of a swimmer in distress: some ignore the cries for help; many help if the situation does not look too dangerous; a few plunge in whatever the risk. The pattern is the same as that of all human characteristics, a spread around some typical value, just as with height or intelligence. As ever, though, it is not the amount of variation that concerns us here as much as the fact that such a response, to provide help to those in need, exists in a general way in the human population.

Why should it be 'good', in terms of the spread of his or her* own genes, for the rescuer to make any effort to save the drowner at all? The crucial point is that in most cases the drowner should have only a small chance of surviving unaided, while the rescuer has only a small chance of drowning. The cost to the altruist is small and the benefit to the recipient is great, just like medical insurance.

In modern society, of course, the two participants in the drama may never see each other again. But for millions of years of human evolution people lived in small groups, tribes and family units. In those circumstances there is a real possibility that if, one day, the former rescuer is in difficulties and needs help, then the rescued man (or woman) will remember the debt owed and will reciprocate. Now, over a period of time, both individuals have achieved a great benefit, at little cost to each of them. In this way any genetic predisposition to help other people out of danger will spread through the population, because altruists will be recognized by other altruists and helped in their turn, while selfish individuals who never help anyone will be less likely to receive assistance when they are in trouble.

It looks almost childishly simple when spelt out like this, because it is so much a part of our nature. It is common sense. But it is common sense only because millions of years of evolution by natural selection have made it part of our nature. And it has taken a special set of circumstances acting on the members of our own species to make us the supreme reciprocal altruists among terrestrial life forms. It does not matter that these requirements are not met by modern people living in citites today: we have lived in cities for only a tiny fraction of evolutionary time. What matters is that the conditions were met by the lifestyle of our ancestors during the past few million years, and that we carry in our cells copies of the genes that made those ancestors so successful.

The first requirement is a long life span, or there will be no time for reciprocation to occur. Secondly, people must live in more or less the same place for most of their lives, interacting

* It is, in fact, much more likely to be a man, for reasons which will become clear in Chapter Eight.

with the same small number of people in a stable social group, where reciprocators and cheats can be identified (not necessarily consciously) and treated accordingly. And, of course, the long period of human infancy, during which the child makes contact with a wide variety of relations over many years, helps the process along. Under those circumstances it is natural that we should have evolved to help each other in times of danger, to share food, to help the sick, to share tools and, perhaps most important of all, to share knowledge. All these activities involve a small cost to the donor and a great benefit to the recipient.

How would the emergence of reciprocal altruism among the groups of primitive humans described by Trivers affect their future evolution? In a burgeoning society of reciprocal altruists, laced with a few selfish cheaters, an evolutionary premium would be placed on many of the qualities that are covered by our modern term 'intelligence'. You need to be able to recognize individuals and assess their reliability; you need some understanding of past, present and future; any move towards more efficient means of communication helps, as it becomes possible to strike genuine bargains, along the lines of 'you lend me your spear today and I'll lend you my axe tomorrow'; and so on. It seems to us that, at the very least, the advantages of intelligence and speech in making us more efficient reciprocal altruists must have been a contributory factor in producing a selection pressure favouring the evolution of those human characteristics. Perhaps this was even a major influence in making us human. R. D. Alexander has summed up the way these pressures would encourage the development of the human brain:

> No feature of the environment is quite so difficult to figure out as what to expect from other social beings with whom we must interact, each of whom is attempting with all of the capabilities he can muster to adjust the outcome of our interactions with him to his own advantage, rather than to ours, when our interests differ.*

The ability to reason, and then to compromise and strike a bargain, stems naturally from the evolution of a complex society of reciprocal altruists, and has now given us the

* R. D. Alexander, *Darwinism and Human Affairs*; quoted by David Barash, *Sociobiology and Behavior*, p. 157.

opportunity to override some of the out-of-date genetic programming that we still carry (as we shall see, especially, in Chapter Eight). We have become something more than instinctive automata responding to the rules laid down in our DNA; the altruistic ape really is different from any other animal on Earth, including the pygmy chimpanzee, in a way which belies the fact that 99 per cent of our DNA is the same as theirs. But still, that altruism is itself evolutionarily advantageous: it is 'selfish' in the appropriate Darwinian sense, it is 'fit'. We can reason out why it is fit, and write books about it; we can refine our natural altruistic instincts into contracts and binding promises; we can work out for ourselves that the pattern of behaviour appropriate to a tribe of pre-Stone-Age people may not be appropriate for a nation armed with nuclear weapons, without (we hope) waiting for natural selection to change our attitudes. But all these abilities exist in people today because they are coded for, directly or indirectly, in our DNA, and because they have proved to be fit – to have survival value – in the past.

Many people, probably most, are reluctant to accept that even acts of self-sacrifice and saintliness are part of our genetic make-up. This is interesting in itself, and some sociobiologists argue that the capacity for self-deception, to convince yourself that you are not really acting to maximize your own success, may be a trait that has been selected for together with the altruistic way of life. We must be careful not to confuse genetic selfishness with everyday selfishness. For an individual, self-*sacrifice* may be a form of genetic selfishness if, as with a bird giving a warning cry, it helps copies of certain genes in other individuals. The bird might not be willing to make the sacrifice if it knew what was going on at the genetic level! Alexander is one who points out that our biology has led to the paradox that we are genetically selfish (like all species) and also dependent on social groups for our continuing success. How can we be both selfish and social at the same time? Why should individuals, in some cases, make the ultimate sacrifice for the good of copies of their genes in the bodies of other individuals? The resolution to the dilemma may be self-deception. The best liars are said to be those who delude themselves into believing their own lies; in the same way, the best social animals – the best altruists – may be those who delude themselves into believing they are acting solely for the good of themselves.

We provide crutches for our altruistic tendencies in the form of morality or religion. We tell ourselves that we should live by a moral code, or by the rules laid down by God, instructions which reinforce the relatively new evolutionary development of reciprocal altruism and help to suppress the older kinds of purely selfish instinct. People are no less human and no less altruistic for the fact that altruism, like all other features of being human, depends on our genetic inheritance. Everything that we are is coded in DNA, and each of us starts out as a single fertilized cell carrying that coded DNA message. But we are not programmed like ants to act blindly in accordance with the statistical rules of relatedness. Instead, we have predispositions that steer us in certain directions. We are inclined to be intelligent, given the right environment and stimuli in infancy and childhood, and we are inclined to be altruistic: to be nice to one another, to love our neighbours. Our feelings of love, affection and friendship are no less real and no less significant for that, just as our feet, say, are no less real and no less important for the fact that their existence too depends on the message that has been passed on as strands of twisted DNA from our parents. Ethics, moral codes and the teachings of the great religions are powerful forces in human affairs because they are right, not in any subjective sense but in the sense that the code of behaviour they represent has been tried and tested in the evolutionary struggle for survival. Peace on Earth, cooperation, helping the sick and weak, and all the rest is a package that has emerged through the process of natural selection. And this makes the case for people to continue to be nice to each other, and to try even harder to be even nicer to even more people, far more compelling than if the message came from a few aberrant individuals struggling to stem the tide of rampant human aggression.

The rules of the game are clear. We are 99 per cent ape, but the 1 per cent advantage lies very largely in the fact that we are altruistic apes. We act out of self-interest, but in most cases this is enlightened self-interest. With this in mind, we can at last understand the basics of human relationships, within the family and in the wide world of politics. And then we will be able to see how best to take advantage of the 1 per cent to make our lives better and more secure.

CHAPTER EIGHT

Subtle Secrets of Sex

'The meaning of sexual reproduction', says Robert Trivers, provides 'the deepest mystery in all biology'.* The central feature of life, as was clear long before Darwin's day, is of course reproduction. Living things reproduce, making copies of themselves (not necessarily exact copies) that continue the line. To a modern biologist, this reproduction is seen in terms of copying DNA and passing on copies of genes down the generations. But why should two individuals be involved in the copying process, with the result that each of them contributes only half the DNA in the new individual? On the face of things it would be much more efficient, in terms of copying DNA, if each individual were to reproduce by some sort of budding process, the way strawberry plants do when they put out runners, creating offspring, new individuals, that contain copies of *all* of the parent's DNA. So how, and why, has sex evolved? And why does it persist in the world?

The only fair answer to these questions, as Trivers's remark suggests, is 'we don't know'. Try though they might, biologists using all the tools of the mathematical approach of the ESS, and all their latest insights into the nature of genetic material and the copying of DNA, still find it very difficult to invent kinds of biological systems in which sexual reproduction has the edge over asexual reproduction. And when they do invent such hypothetical systems, these bear little or no resemblance to the world. The reason is simply the doubled efficiency of the asexual method in terms of copying DNA. An individual that has to mate with another individual in order to produce offspring at all suffers many disadvantages in life, from the evolutionary point of view, because of the need to find a mate, go through courtship rituals, run the risk of being surprised by a

* Robert Trivers, *Social Evolution,* p. 315.

predator while in the act of mating, and so on. But the overriding factor in all these calculations is simply that if only half of your DNA is passed on to each of your descendants, then you have to be involved in 'making' at least two descendants in order to ensure that copies of all of your genes are passed on to the next generation. All the calculations show that in species as diverse as ourselves and flowers, frogs and antelope, an asexual mutant that suddenly appeared in the population – a female that gave birth to exact genetic copies of herself, without her eggs being fertilized by sperm from a male – would be so successful that her descendants would wipe out her sexually reproducing relations within a few generations of intense competition. Indeed, exactly this process has happened in some plants, such as the dandelion, which have abandoned sex and reproduce in precisely this way, sending out seeds that are fertile without ever having been fertilized. And the dandelion is a very successful plant.

Evidence such as this, together with the mathematical calculations, suggests that today there may well be no evolutionary advantage to sex. Sexual reproduction may indeed have been very useful long ago, so useful that some ancestral species which practised, or invented, sex has given rise to the enormous diversity of sexually reproducing species on Earth today, but the conditions that once favoured sex have now disappeared. In this picture, sex is a hangover from an earlier era of evolution, so that sexually reproducing species are indeed at the mercy of asexual mutants. But whereas this kind of mutation can occur relatively easily in a plant (explaining the success of dandelions today), it is almost impossible in a mammal, such as ourselves, because the machinery of sexual reproduction has become so complicated. This is the basis of the claim that men are, in strict evolutionary terms, redundant today, if only women could find a way to follow the success of the dandelion in producing fertile, but unfertilized, eggs.

We are not, however, particularly concerned with the origins of sex in this book. Intriguing though the question is, it is one we shall leave for the experts to puzzle over. What we are concerned with is the fact that people do reproduce sexually, with the contribution of both a man and a woman being required to produce a new human being. That basic genetic requirement is the single most important influence (together with the long infancy of the resulting baby) on human society and human behaviour. Even the most blinkered opponent of the ideas of sociobiology cannot fail to see that there are men and women in the world, and that this biological fact has social implications! Before we look at the implications, though, perhaps we ought to get the basic biology straight.

Sources of sex

Sending copies of your genes on into succeeding generations is what life and evolution are all about. Successful genes, ones that make their carriers more fit, in Darwinian terms, squeeze out unsuccessful genes. In human terms, an adult female who has survived to reproductive age is obviously successful; daughters who are exact copies of herself ('clones') ought to be just as successful, so why bother mixing in a 50 per cent proportion of genes from a man?* The most obvious result of this mixing is that it produces variability. Instead of identical daughters (identical to each other and to herself), the mother produces children that carry new mixes of genetic material and who differ in different ways from each other and from herself and their father(s). Variability is so obviously the key feature of sexual reproduction that it must be intimately related to the origins of sex, and to the success of sex in the past. Variability can be an advantage in itself, under special conditions when the environment is changing, and the mixing of genetic material involved in sexual reproduction can help advantageous mutations to spread through a population. But these advantages are usually small and outweigh the twofold advantage of asexual reproduction only under special circumstances, circumstances that probably do not apply to the vast majority of species on Earth today, and certainly do not apply to ourselves.

In everyday language, sex speeds up evolution. But that is an

* Strictly speaking, a clone is a *group* of genetically identical individuals. The individuals are therefore *members of* the clone, in the correct but cumbersome terminology, but in recent years popular usage has altered the meaning of the word, so that often nowadays you see reference to 'a clone' meaning an *individual* that is an exact genetic copy of another individual. As good Darwinians, we follow this evolved usage of the term.

advantage only where there is some need for rapid evolution. That might have been true in the past, when conditions were (perhaps) more variable than they are today. Or it might be something that is true from time to time. As we have seen, fossil evidence shows that there have been repeated occasions in the history of our planet when there were massive extinctions of many life forms. These disasters – such as the 'death of the dinosaurs' some 65 million years ago – may have been linked with natural phenomena such as ice ages, or with the effects of the impacts of giant meteorites with the Earth. In the aftermath of such a disaster, sexual reproduction would be an advantage and might result in species which 'use' sex becoming so well established that by the time things had settled down into a more stable state there was no immediate scope for asexual mutations to take over. If a similar disaster struck tomorrow, the dandelions might well disappear, their ecological niche being taken over as the environment recovered by some variation of sexually reproducing plant that survived the holocaust and spreading out into all available niches. But in several million years time an asexual mutation might once again crop up to dominate the now-stable ecological situation.

The best prospect of explaining why rapid evolution might be an advantage, however, comes from looking at how predators and disease affect a population of animals. Some of the latest ideas about the role of sex in evolution concern these links, especially those between large creatures like ourselves and the tiny organisms, bacteria and viruses, that cause disease. These microorganisms have a very rapid life cycle, and reproduce many times in a few minutes, let alone in a human lifetime. Partly for this reason, they are constantly evolving new variations on old themes: different strains of influenza, for example, crop up every few years, and sweep around the globe. Somehow the defences of the body have to evolve rapidly enough to counter these changes in the invading organisms, and there has to be a variety of individuals, otherwise the defences will be overwhelmed and all the identical members of the species that is being attacked – the ones that get ill – will be wiped out. It is just possible that sexual reproduction survives in the world today because it provides enough variability, and rapid enough evolution of defences against disease, to ensure that we and species like us are not wiped out by super-bugs.

This is ironic, because we are all descended from microorganisms very similar to the bacteria that are so potentially lethal to us today. Sex got started when pairs of those primeval bacteria-type cells 'learned' how to get together and mix their genetic material, instead of just splitting into two new cells each containing an exact copy of the parent's DNA. Again, nobody can be sure how such a system developed, or what the initial advantages were that gave it a leg up the evolutionary ladder. One suggestion is that by combining two copies of the primitive genetic material such early cells were able to eliminate errors. A mistake in a stretch of genetic material from one cell, which might be potentially lethal, could be rectified by taking on board a correct copy from another cell. Whatever the exact origins of the sharing process, however, it does seem that once it got started there was an inevitable tendency for a polarization which has led, down the eons, to the two sexes – and two sexual roles – of the modern world.

The best guess we can make about the origins of sex is that it involved, for whatever reason, pairs of more or less identical cells coming together, sharing their genetic material, and then dividing into two (or more) daughter cells which each carried a different mix of genetic material than their parents' mix. Such a system is sexual reproduction all right, but it requires only one sex as all the cells are essentially the same type. But it does not take the mathematical wizardry of John Maynard Smith to see how evolutionary pressures will drive the descendants of these cells in two different directions.

There are just two ways to be successful, in terms of producing viable offspring, under such a system. It is absolutely essential to find a partner, once such a system has become established, or you will not be able to reproduce at all. And it is no less essential to provide a good store of raw material ('food'), so that each of the daughter cells is big enough to survive and function efficiently. These requirements are to a large extent opposed to one another. A small active cell will be able to swim around rapidly in the primeval ooze and find a partner, but it will not be able to carry much in the way of resources for the next generation. A big fat cell will be an excellent provider of resources, but will not be very mobile, and will have to wait for a 'mate' to come its way. Small cells that meet one another and mate will produce even smaller offspring, ripe for being gobbled up by other microorganisms; big fat cells will not find a mate at all, unless a small, mobile cell swims up to them. Immediately, we have the beginnings of the sexual system of reproduction involving two sexes: a 'female' that provides a large 'egg', and a 'male' that provides nothing more than a package of genetic material, a 'proto-sperm'.

This is, of course, a great oversimplification of a complicated process. But it does suggest how two sexes could have emerged very early in the story of life on Earth, and it underpins the most basic feature of sex roles in human beings and many other species: the female provides a large egg, and the male provides many small sperm. This does not necessarily mean that the female has to be the nurturant partner, taking care of the babies. In some species of water bug, for example, the male carries the fertilized eggs around on his back, looking after them until they hatch. Women today are commited to carrying their offspring in the womb for nine months, but many men are able and willing to take over some of the responsibility after birth. But to a biologist the fundamental truth is that the female of any species is the one that provides the larger sex cell, or gamete – in everyday language, the egg. And the resulting differences can be crucial for the behaviour of the individual male and female members of the species, each individual being concerned with maximizing its own evolutionary success by producing as many offspring as possible.

Sex roles and sex ratios

Before we proceed any further, we should stress that identifying features of human behaviour and the reasons why such features have evolved is quite different from the question of whether we 'ought' to behave in a certain way. Millions of years of evolution have clearly produced women adapted to looking after babies. Leaving aside pregnancy itself, only women have breasts that produce milk for babies to feed on, and until only a few decades ago it was impossible for any man to take complete charge of rearing an unweaned baby. That is an evolutionary fact. But today, of course, a man can take a newborn infant and look after it entirely on his own, without any help from a woman, because he can obtain milk formulas derived from cow's milk to keep the baby healthy and happy. We are able to override our evolutionary inheritance through a combination of intelligence, technology and social adaptations. But it is still useful to know the underlying biological rules of the game.

In a similar way, our upper bodies, especially our arms and shoulders, are adapted for swinging through trees from branch to branch: the form of locomotion called brachiation, as we saw in Chapter Two. The genetic and evolutionary origins of this are clear, but that does not mean that we all use the skills of the gymnast in our daily lives. So when we come to discover that human evolution during the past few million years has produced males with an inbuilt tendency to seek more than one sexual partner, that does not mean that we are suggesting that all men 'ought' to be philanderers, any more than we say that all people ought to swing through the trees, or that no man ought to feed a baby. Intelligence brings with it a new understanding, new types of society (themselves biologically determined, of course, since intelligence is a biological phenomenon) and new patterns of behaviour. But if we understand the genetic imperatives that undoubtedly do incline men and women to have different attitudes towards, in this case, sex, then we shall understand better how society should take care to encourage people, where appropriate, to override those imperatives and to act in ways that are suitable for survival in the civilized world today, as opposed to the African savannah 3 million years ago.

Reproduction is very closely related to evolutionary success, and different strategies of reproduction are particularly susceptible to natural selection. As David Barash points out,* the result is that 'the analysis of reproductive strategies has been one of the most productive areas of sociobiology'. We do not have the space to go into all the examples of this productive line of research here, and shall concentrate chiefly on the examples most relevant to ourselves as human beings. It happens that humans are rather unusual, as mammals go, and that our

* David Barash, *Sociobiology and Behavior*, p. 14.

unusual features are largely a result of the long-term investment parents have to make in bringing up children (something we delve into in more detail in the next chapter). The human pattern of mating is much more like that of birds than mammals, and for the same basic reason: birds too have to invest a lot of effort in rearing their young to the point where they can begin to lead independent lives. Very many mammals go in for a quite different kind of mating system to the human one. Ours, of course, is dominated by partnerships between one male and one female which last for a long time. The more 'natural' system for many mammals is to have a harem of females dominated by one or a few males. This causes hackles to rise on some people, who mistakenly believe that sociobiologists who study such systems are suggesting that human males 'ought' to dominate harems of women. But this type of mating system is even more interesting because it raises a question that completely baffled Darwin himself, but which we can now answer with great ease.

Let us take the puzzle first. In a species like our own, the females are each physically capable of producing only a very limited number of offspring, compared with the reproductive potential of an individual male. We can put the figures in perspective by considering the human case itself. Even allowing for the possibility of multiple births, it is extremely unlikely that any woman could produce more than 50 children that survived to reproduce in their turn. A much more realistic figure, in line with known records of women who have proved particularly fertile, would be half that number, say 25. This is simply because it takes nine months from conception to birth, during which time, of course, there can be no further conception; indeed, it is unlikely that the woman will conceive again within three months of the birth, giving a maximum 'productivity' of a child per year. A man, on the other hand, if healthy and in his prime years of sexual activity, could quite possibly mate with a different woman every day of the year if he could find enough willing partners. This is extremely unlikely, but it makes our point. But a man could easily become the father of as many children in the space of a couple of years as a woman can produce in her entire lifetime; and the record for male productivity (amongst the harems of old-time eastern potentates) runs into several hundred children fathered in a lifetime.

It might seem from this simple fact of life that there is no 'need' for very many males in the population. Many females are necessary to produce the next generation of individuals, but surely a few males would be sufficient to do the job of fathering them? In fact, this is a fallacious, group-selectionist argument. For individuals, the best strategy (except under very special circumstances) is to become the parent of an equal number of sons and daughters.

Darwin puzzled over the phenomenon that is very many species, including our own, the ratio of males to females is very close to 1:1. He could not see why it should make any difference in evolutionary terms whether an individual left behind all sons, or all daughters, or some arbitrary mixture of each. The problem was solved in the 1930s, by the mathematician Ronald Aylmer Fisher, simply by considering the next generation of descendants: the grandchildren of the individuals you start out with.

The argument runs like this. Suppose there is a species in which there is a genetic predisposition for females to be born, so that in the first generation of descendants of our hypothetical parents there are three times as many females as males. It follows immediately that those males will, on average, be three times more successful than the females in passing on their genes to the next generation. At one extreme, if the individuals form male–female pairs and are monogamous, two-thirds of the females will be without partners and have no offspring; at the other extreme, if the individuals are completely promiscuous, all the females will be inseminated and will have offspring, but on average each male will have inseminated three females, so each male will pass on three times as many copies of his genes as each female to the next generation. As long as the population is biased towards producing females, any mutation which causes one individual to produce sons instead of daughters will do well because those sons' genes will in turn spread three times as effectively as female genes, and carry with them the mutation for maleness.

Exactly the same argument applies in reverse, for an original hypothetical population containing three times as many males as females. In that case, females will have greater evolutionary success, and any mutation encouraging the production of females will spread rapidly through the population. The only stable strategy, the ESS, is for the populations of males and females to remain in balance, with a ratio of 1:1.

This is true even among species that practice polygamy. To take another hypothetical example, suppose there is a species in which, on average, each successful male has a harem of ten females that he dominates and from which he excludes all other males. You might think at first that the correct strategy for a mother 'ought' to be to give birth to ten daughters for every son, since any additional sons will be wasted, in terms of evolution, and leave no descendants. But remember that each successful male inseminates ten females. He has a one-in-ten chance of winning a harem, and his payoff is at odds of ten to one, exactly matching the chances of success. The net result in that the 'correct' ESS is still for each female to give birth to equal numbers, on average, of sons and daughters. All her daughters will reproduce, and the few sons that achieve any sexual success will hit the jackpot and produce ten times as many offspring as any one daughter.

The 1:1 ratio holds almost universally, not just among mammals but among all vertebrates. There are exceptions, though, and these have provided some of the neatest tests of sociobiology, confirming the accuracy of the ESS approach. Ants, for example, have the same kind of relatedness pattern that bees have, with female workers sharing 3/4 of their genes with their sisters and only 1/4 with the brothers. In such a situation, if you follow through the same line of argument that Fisher pioneered over fifty years ago, a nest in which the worker ants dominate the raising of the infant ants 'ought' to raise three times as many sisters as brothers, because only when the ratio of the sexes is 3:1 does the greater reproductive success of each male balance the fact that he shares only 1/3 as many genes with his sisters as they do with each other. This was a classic example of a scientific prediction, made by Trivers and his colleague Hope Hare. They made the calculation first, then counted the numbers of males and females in many ant colonies. Lo and behold, in all the species for which this kind of relatedness holds, the ratio is indeed 3:1 And that kind of successful prediction helps to instil confidence in other applications of the ESS approach, and in sociobiology in general.

One of the big questions about human sex and reproduction is why we should usually pair up with one partner for the purpose of raising children. A romantic will tell you that it is because we fall in love, and that love is for ever. But of course, love is something that has evolved over countless generations as a form of pair bonding. It has evolved, and persists, because it has proved itself a success in evolutionary terms: that is, people who pair up and raise children together in a stable relationship have, in the past, produced more children that live to reproduce in their turn than have people who 'chose' other forms of reproductive partnership. Any inherited tendency for people to 'fall in love' will therefore spread through the population.

Perhaps we can make our point more emphatically with an even more clear-cut example. Most people enjoy sexual intercourse, with the right partner (although some seem not to care too much about the partner). Maybe our remote ancestors did not particularly enjoy the act, but carried it out instinctively, as many animals seem to. Some, including our closest relations – the pygmy chimpanzee in particular – obviously *do* enjoy sex; others, such as honeybees or fish, seem to act purely on instinct. Once again, we see human behaviour carrying a characteristic shared with our nearest relations to an extreme. With the development of our large brains, we have developed a greater capacity for pleasure, and any individuals that happened to get pleasure out of sex would be more likely than others, for obvious reasons, to leave many offspring, some of which carried that same capacity for pleasure. And, of course, any individuals who actively *dis*like sex are unlikely to leave large numbers of descendants!

Why monogamy?

The difference between the sexes is expressed very fully in mammals. Although both parents contribute their genetic material to the fertilized egg, at base level that is *all* the male needs to contribute: a few minutes of his time, a small amount of sperm. The female, on the other hand, is committed to weeks or months spent carrying the developing fetus inside her, and then to a further period in which the offspring is dependent on her for milk, even if it is able (as with deer for example) to run with the herd almost as soon as it is born. As a general rule, females have a bigger investment in their offspring; males have little or no investment. Because females do nearly all the work of reproduction, they are in effect a resource, one for which males may compete with one another. In our hypothetical example only one in ten males get to be fathers at all, so the competition is fierce, and the rewards great. But there are natural systems, in species alive on Earth today, in which competition is even fiercer.

In most species the members of the two sexes also show marked differences in the way they choose a sexual partner. Females tend to be choosy, while males are not. Usually each female has a choice of mates, and picks one by a process that is certainly not random; even in harems, which males win by fighting, the females are choosing to bestow their favours on the most successful fighting male. With most females using the same criteria for selection, in many species just a few males get most of the opportunities to mate. Many of the rest may miss out completely. Among males, winner takes all, and in the reproductive stakes most win nothing. So males tend to be indiscriminate in their sexual activity. It requires so little effort of them that, by leaping at any opportunity to mate, they can make sure that a few successes – matings which result in the production of fit offspring – more than make up for the time they waste when they choose an unsuitable partner. And some choices of partner can be bizarrely unsuitable: male bullfrogs, for example, will grasp in the mating posture almost anything that is roughly the size of a female bullfrog, and that includes the foot of a Wellington boot worn by someone tramping through the swamp where the frogs live! The gerbils of the Sahara Desert provide another example of how the male will risk his life for an opportunity to mate, while the female is more cautious. Individual females seek out small regions where there is a little plant life to eat, and defend them against invaders; their distribution is related to the distribution of food. But the male is driven as much by the need to mate as the need to eat. Each male will seek a site to dig his burrow which is as close as possible to as many females as possible, even if that means there is little food. The distribution of males is determined by the distribution of females. Not only does he make do with a poor diet, the male repeatedly risks death by running across hundreds of metres of open country to visit a female, on the off chance that she may be willing to mate with him before she chases him out of her precious territory. The whole business is so dangerous that the entire adult male population of an area may be replaced every few months, whereas a female has a good chance of living for a year or more, in spite of the effort she has to put in to raising her young.*

The biology of females makes them careful, literally putting all their eggs in one evolutionary basket, and having a few offspring that they must raise carefully, for which they require the fittest mate they can obtain. The biology of males makes them live dangerously, to be philanderers, to mate indiscriminately and not give a damn what happens to offspring they may never see, on the basis that if there are enough of those offspring some are bound to survive. The rule holds throughout the animal kingdom, so why should we be different, if indeed we are different.

European red deer, which have been studied in great detail by Cambridge University researcher Tim Clutton-Brock and his colleagues, provide an archetypal example of the mammalian mating strategy, and how it is different for males and females. The male deer, the stags, are bigger than the female deer, the hinds, and have impressive antlers which they use in fights with other males. The biggest stags, with the most impressive antlers, generally win these fights, which is why the species has evolved in this direction. A successful stag may hold a harem of as many as twenty hinds, from which he excludes all other stags, so that he is the father of all the offspring born to those hinds. He achieves a very high reproductive success in the short term, directly as a result of his size, his antlers and his competitiveness, fathering twenty or more offspring in the course of two or three years. However, he is soon exhausted by his efforts to maintain his harem and to exclude other stags, and will be defeated, sooner rather than later, by a younger, fitter stag which then takes over the harem. And while a few

* Martin Daly and Margo Wilson, *Sex, Evolution, and Behavior*, p. 73.

stags father more than a score of offspring each, most stags leave no offspring behind at all.

The situation for the hinds is quite different. Virtually every female that lives to breeding age will breed, but each of them can produce only one offspring in a season. Even the most successful hind will produce only a dozen or so offspring in her lifetime, perhaps half as many as the most successful stag. But she will not have the stress of fighting, and she will almost certainly live longer than the successful male. And these lifestyles are implicit in the mammalian method of reproduction, as we have seen. Males 'should' be inclined towards polygamy, because they have much to gain by this system of reproduction if they are successful. To the female, it makes little or no difference whether she is polygamous, as long as the father of her offspring is an evolutionarily fit male. (Of course, it is an advantage for a female if the father of her offspring is a big, successful male since it is likely that he will pass on these attributes to her sons, who will grow up to be big, successful males and father many offspring in their turn, carrying copies of her own genes on with theirs into succeeding generations. So it is no mystery that evolution should have selected females that 'chose' to stay in the harems.)

This last point is strikingly borne out by the behaviour sometimes observed in lion prides. These groups are built around a core of breeding females, together with two or three adult males and the young of the pride. Young males are thrown out of the pride before they become sexually mature, so the dominant males father all the offspring, until they are ousted in their turn by outsiders – perhaps youngsters ejected by another pride – as the dominant males become older and weaker. When that happen, the 'new' males immediately kill all the cubs in the pride, and begin an almost frantic round of mating with the females. From the males' 'point of view', this is logical. The cubs they have inherited are no kin of theirs, and do not carry the genes of the new males. In terms of the survival of their own genes, the right thing to do is to get rid of these unwelcome nuisances, and get the females pregnant with their own offspring. Removing the cubs has a twofold advantages: it destroys the offspring of what are rival lines, in terms of evolution, and it causes the femals to come into oestrus. But why should the females 'allow' such behaviour? Surely it is in their interest to see their own cubs survive?

This obvious truth is only half the story. The females 'want' their cubs to survive, but it really does not matter who the father of those cubs is, provided he has proved his fitness. Presented with a *fait accompli* – dead cubs – the best strategy for the female is to mate with the dominant male(s) as soon as possible and produce a new cub that will be raised to maturity by the pride. And (although perhaps at rather too subtle a level to have much effect on the evolution of this system) the 'new' father must be 'fitter', in some sense, than the old one, since he has just won a takeover battle with the previous dominant males. (We do not wish to labour the point, but a little thought along the lines indicated by the lions' example provides very interesting insights into the stories of wicked stepfathers and wicked stepmothers immortalized in so many legends and fairy stories, such as Cinderella, which clearly echo some deep human experience.)

It is easy to see why polygamy, and especially polygyny (in which one male mates with several females) is the right evolutionary strategy for mammals. Biologists are much more interested in the exceptional cases in which mammals are monogamous, especially in view of the bearing that any insight into those special cases may have on our own way of life. One very neat approach to this problem has been made by biologists, such as Burney Le Boeuf of the University of California at Santa Cruz, who have studied closely related species which share similar lifestyles but have different mating systems. La Boeuf has paid particular attention to the elephant seals, where dominant males control enormous harems and father very many offspring, and other pinnipeds, such as crabeater seals and harp seals, that are monogamous. Elephant seals are, in any case, interesting as an example of an *extremely* polygynous species. The males, up to 4m long and weighing in at 3000 kg, are as much as five times larger than the females, and this is interpreted as a direct result of the fierce competition between males in their efforts to become owners of harems. Throughout the animal kingdom, this kind of difference in size between the sexes – called sexual dimorphism – is associated with polygyny, and it is a reliable rule of thumb that the greater the difference in size between males and females, the bigger the size of the harem that a successful male will dominate (and consequently the greater number of males that never get to breed at all). The higher the reproductive stakes, the bigger the male of the species. Elephant seals mate on land, on the beaches of islands off the Pacific coasts of California and Mexico, where hundreds of females lie packed together like sardines in a tin, and where males take part in prolonged battles which end with a few of them gaining the 'right' to mate with an enormous harem.

There are 34 members of the order Pinnipedia to which elephant seals belong; 21 of them breed on land, in similarly restricted sites, crowded onto beaches or narrow sandbars. All but three of those are highly polygynous. Although less is known about the 13 species that breed on ice, most of them are clearly monogamous, and during the breeding season they are seen in isolated pairs, or in 'triads', a mother and her year-old offspring, plus her mate. Where the females are crowded together, the archetypal male strategy of polygyny is allowed to evolve to extremes. But on the ice females do not crowd together. Sharks, killer whales and polar bears all prey on pinnipeds, and a crowd of seals huddled together on the ice would be extremely vulnerable to predators. Indeed, it was probably the predators that drove the ancestors of the elephant seals and other shore breeders out of the sea to give birth. This in turn led to crowding on the few available islands, and the rise and rise of polygyny. Ice breeders followed a different tack. First, there is much more ice than there is safe beach, so there is room for females to spread out, each well able to obtain food from the sea nearby. In addition, the ice breaks up as spring progresses, so there is a limited time during which it is safe to give birth, and almost all the females do indeed give birth at the same time, which means they must all have mated at the same time. A spread-out population of females, each of which needs to find a mate at the same time each season, inevitably implies that there is one male mating with each female.

Exceptions to these simple rules, such as species that breed on ice held fast to the coast of a landmass, have breeding strategies that are also explained by the special requirements of their environments. The details need not concern us; what we are interested in establishing here is that although polygamy, and especially polygyny, is the basic reproductive strategy of mammals, there are special circumstances in which this basic strategy has been altered by the environment to produce other breeding systems, including fairly strict monogamy. When a sociobiologist says that polygamy is the natural strategy for mammals, that is not the same as saying that polygamy is the natural strategy for people. True, we are mammals, but like the crabeater seals, we are a special case. We are *almost* monogamous: it does seem that we still carry traces of a polygynous past, part of our basic mammalian inheritance.

Monogamy is more common in birds than in any other group of animals, and is the rule in more than 95 per cent of all known bird species. As we have seen, this is related to the enormous effort that parents have to make to raise their chicks successfully, an effort that requires the full-time attention of both parents. Although there are species in which various forms of polygamy operate, and there are species in which males try to take advantage of opportunities to mate with females while the female's partner is otherwise occupied, by and large monogamy rules the roost. This is reflected in the similarity between the two sexes in such species: a male swan, for example, differs little in size or even plumage from a female swan, and even in species where the male goes in for gaudy plumage in order to attract a mate, there is little difference in body size between males and females because the males do not physically fight one another for control of a harem.

Starting out from a typical mammalian mode of reproduction, human beings have gone a long way down the same path as that taken by birds, and for the same reason: the need for both parents to cooperate in raising the young, if the young are to survive to maturity and reproduce in their turn. But we have not yet gone as far down that path as the vast majority of bird species.

Unlike so many bird species, the human species is characterized by a sexual dimorphism in body size. Men are, on average, about 8 per cent bigger than women. In any other species, this would immediately suggest to a biologist studying the species that there is competition between males for females, and that the most successful males 'ought' to have two or three wives, while many have only one wife and a significant number fail to find a mate at all. This clearly is not true of modern Western society, where monogamy is the rule (or at least where one man and one woman are partners for a time, even if the partnership is later broken up by divorce and both partners then form new monogamous relationships; it is no coincidence that divorce occurs most commonly after the children of a marriage are big enough to stand on their own feet). So is the theory that sexual dimorphism is related to polygyny wrong? Probably not.

What the difference in size between men and women really tells us is that there has been a selection pressure favouring slight polygny, at least until very recently during human evolution. Even if conditions in the civilized world today do not favour polygyny, we still carry the genes for males to be bigger than females, and to compete with one another. It will take many generations for the tendency towards sexual dimorphism to be diluted and washed away. Of course, it is easy to see how in primitive society these differences could have arisen. Two women at home looking after the children, with one man the father of those children, keeping off potential rivals, makes a family unit that is recognizably one with survival value when it comes to rearing human infants. We are not suggesting, however, that the poor, helpless women needed the man to protect them, and to provide food by hunting. Most studies of 'primitive' societies show that women provide most of food, by gathering, and that men are able to contribute only a little by hunting. In our hypothetical family group, the women 'need' the man primarily as a father for their children, and to provide some assistance with obtaining food. We do, of course, accept (one of us reluctantly) that a woman in the late stages of pregnancy would need help from other human beings in order to find food and survive. The fact that this other human being might very well be another woman, however, probably tells us more about the basic unit of two women and one man than does the myth of man the great protector. Two or three women in a family group could feed themselves and their growing children, tolerating the existence of a man who they need to father the infants but who they throw out, as often as they can, to amuse himself with hunting or with fighting other men.

As society has developed, two new evolutionary pressures

worked against this family unit. First, one man and one woman – or, perhaps we should say, one woman on her own – have become well able to raise several children together, thanks to inventions such as agriculture and technology. Secondly, society as a whole is now able to take on all or part of the burden of child-rearing, through crèches, nursery schools, and schools proper, releasing women almost entirely from what has been their primary biological function since the mammalian line began. Some of these changes happened thousands of years ago, others only within the past few centuries and decades, a mere eyeblink in evolutionary time. Small wonder, then, that even the most civilized of us still carry genes that were selected for under quite different conditions to those in which we live, and small wonder that many societies in the world today still conform more closely to the mammalian stereotype than those of us who live in cities such as London and New York can appreciate. Researchers Martin Daly and Margo Wilson, in their book *Sex, Evolution, and Behavior*, list several examples that bear this out. In general terms, among all the human societies they investigated, polygynous marriage is common, while polyandry (one woman taking several husbands) is rare. Out of 849 societies investigated, they say, polygyny occurs in 708. It is common in roughly half that total, and occasional in the other half. Monogamy is the common pattern in just 137 societies, and polyandry is found in just 4, all of them special cases where biologists can see reasons for the system being practised.

Similar studies highlight other features of human relationships. Whereas in most mammalian species sexual relations are largely impersonal, and alliances between males and females are at best temporary, there is no human society in which sexual relations are casual and impersonal. Marriage of some kind is universal, in all cultures; but, as we would expect of mammals that are being pushed into monogamy, bachelors are more likely to be found than spinsters. Although a majority of societies allow polygyny, the majority of marriages are monogamous, because even in societies where polygyny occurs it is only the most powerful or wealthy men that take more than one wife at a time. And when it comes to choosing a mate, both men and women are as predictable in their behaviour as the males and females of other species.

Choosing a mate

The way in which marriage is institutionalized by society shows how intelligently our ancestors coped with the sociobiological implications of our genetic inheritance, long before genes and sociobiology were thought of. As a social way of life emerged, it must have been clear to the intelligent leaders of tribes (or villages, or whatever units of society) that young, unattached males are troublemakers. They fight among themselves, chase after the girls and generally carry on in a disruptive manner. But, by and large, males that have settled down with a mate are good, hardworking members of society. With the sexes roughly in balance numerically, the best way to settle everyone down is to marry off each young male with one female, maximizing the calming influence of matrimony. This is a good thing for society, but runs counter to our genetic inheritance, and this is why marriage is built around with rules and customs, and became a great institution. If it were as natural for human beings to pair off for life as it is for swans, there would be no need for the institution of marriage at all. This is just the kind of insight sociobiology can provide. Hopefully further sociobiological insights will in future help us to tailor our institutions even more accurately to the needs of society, and of individuals within society.

One rather good example is provided by the attitude of society to incest, which represents a rather unusual form of choosing a mate, but which holds enough fascination for many people to show that it has a special place in the evolutionary story. There are good biological reasons why individuals should prefer, by and large, not to mate with their brothers or sisters, and ways of avoiding incestuous relationships have evolved in most vertebrate species. The reasons why brother–sister matings are often bad things, in terms of evolution, is that if the brother and sister both carry copies of a defective gene, then there is an increased risk that their children will inherit that allele on both the relevant chromosomes, so that the child will not develop properly and may be deformed or sickly, or die young. As a result, natural selection will favour individuals that choose, for whatever reason, not to mate with their siblings. Such outbreeding individuals will, by and large, tend to have healthier and, in the Darwinian sense, fitter offspring.

The evolutionary mechanisms for avoiding incest have evolved many times and operate in different ways. The way lions expel male cubs from the pride before they reach sexual maturity is one obvious example, which we particularly like because it shows that a social mechanism may be selected for by more than one evolutionary pressure. The dominant males expel potential rivals and remove the risk of incest, killing two evolutionary birds with one stone. We do not say that the ancestors of today's dominant males started to expel youngsters in order to avoid incest, but once this behaviour is established as an ESS it will be reinforced because it promotes outbreeding (as long, of course, as outbreeding is a 'good thing' in its own right). In other species, such as mice and rats, individuals are reluctant to mate with their litter-mates, who they may well be able to identify by smell. It is literally true, in many species where several young are reared together, that familiarity breeds contempt, in sexual terms, and our own species also seems to operate like this.

Children that are reared together and play together in unrestrained fashion almost never become marriage partners in later life. This is shown most clearly by the example of the Israeli kibbutz system, where youngsters from different biological parents are reared together as if members of one big family. In spite of the often deep wishes of their parents for young men and women from the same kibbutz to marry one another, there is no recorded case of this ever happening. The human incest-avoidance system, whatever it may be, operates on the basis that the children you are brought up with should not be sexually attractive to you in adult life, rather than, say, by sense of smell. This is especially interesting because in prudish 'Victorian' societies boys and girls are kept apart and not allowed to play together, even within the same biological family. And this can have exactly the opposite of the intended effect, since in adult life the young man and woman, although biologically siblings, have never received the 'message' that they should not regard each other as potential mates!

But there is even more relevance in all this to society today. Thanks to modern contraceptive methods, there is no reason why a brother and sister should not, if they wish, choose to be sexual partners. And thanks to modern techniques in genetic counselling and genetic engineering, there is no longer even any evolutionary reason why they should not have normal, healthy children. The incest taboo arose in society in the first place because the way in which children were being reared had removed, at least partially, nature's instinctive incest-avoidance mechanism, but further changes in society have made the taboo no longer necessary. Here is a clear-cut case where sociobiological insight can help us to change the rules of society (as has been done in Sweden, for example) to relieve the misery of some of its members.

Apart from avoiding your immediate kin, however, the human preference when seeking a mate seems to be to grab the nearest one that is free. This rather mundane conclusion is the result of a wealth of research carried out by David Buss, of Harvard University. People tend to find marriage partners among the people they see most often: their neighbours at home, at work or at college. One of the slightly less obvious conclusions Buss's work has thrown up, however, is that there is no truth in the old saw that 'opposites attract'. In fact, people tend to marry people who are as much like themselves as possible, in intelligence, social class and even appearance, a tendency sufficiently pronounced that some researchers use the term 'associative narcissim' to describe it. There are, though, some differences between what women and men look for in a mate. In surveys, women consistently indicate a greater preference than men for a partner who has good earning capacity, comes from a good family background, has professional status and is kind and gentle. The attributes in a life partner that men value more highly than women include physical attractiveness, frugality and being a good housekeeper. These requirements conform so clearly to the differences between male and female mammalian reproductive strategies that there is no need for us to highlight them further, except in one regard.

In terms of evolutionary success, the prime requirements that a man should seek in a marriage partner is a combination of youth and health. A young, healthy woman will be able to produce many children for him, so over countless generations men who prefer to mate with young, healthy women left more descendants than men who preferred older woman. In that sense, it is natural that men today, descended from many generations of such matings, should be most strongly attracted to young women, just past puberty. A woman, on the other hand, required rather different attributes in a partner, until very recently. She needed a male that had proved himself fit, like the red deer stags. And a man proves himself fit firstly by surviving to a reasonable age, and secondly by achieving wealth or status within society. A woman who mated with a successful older man was likely to leave more descendants, in the long run, than a woman who chose a young, unproven male as her partner.

Although the reasons why partnerships between older men and younger women should have proved evolutionary success in the past no longer apply in the same way in modern society, this insight enables us to understand why it is still true that older men are attracted by younger women, and vice versa. Of course there are exceptions: as in all our examples, we are talking only about statistical rules, which apply by and large throughout the community. Some young men prefer older women as sexual partners. But they are a minority, and do not weaken the main thrust of the argument.

This particular insight can again tell us something useful about things we are doing wrong in society today. Advertisers who use very young, pre-pubescent girls dressed up in adult clothes and make-up to promote their products have recently come under attack from pressure groups who suggest that this kind of advertising encourages some men to treat children as sex objects, and may be directly linked to a rise in the number of cases of sexual abuse of children. The insight provided by sociobiology strongly supports this contention. Powerful forces have moulded human males over a long time so that their sexual urge should be directed at young women. It is no surprise that by presenting girls that have not yet reached puberty in a 'disguise' which makes them look more mature it is possible to cause this natural mechanism to overreact, with disastrous consequences for some individuals. But the advertising would not produce this reaction – indeed, it would not exist – if men were not selected to find young women particularly attractive. Once again, the majority of men will not be affected in this way, certainly not enough to lead to direct attacks on young girls. But just as a few men are attracted to older women, so a few may already find extremely young women, or girls, more attractive than others. Society should take great care not to confuse such individuals further, and in sociobiological terms we regard the case against this kind of advertising as proven.

The phenomenon of like attracting like, however, also sheds new insights on the recent evolution of *Homo sapiens*. This process, called 'assortative mating' in the jargon, tends to increase the frequency of occurrence of combinations of genes (genotypes) that produce extreme phenotypes. At the same time it decreases the frequency of genotypes that produce more average phenotypes. Take height as an example. Tall men tend to marry tall women, and short people also tend to pair up. So short people have short offspring, while tall people rear tall children, whereas if each tall man married a short woman, and vice versa, there would be a tendency in the next generation for height to even up.

Small effects like this can produce significant changes over many generations, increasing the differences between people. And since people today are more mobile than ever before, meeting many more other people in their lives than our ancestors ever did, there is more opportunity than ever before for individuals to pair up with other individuals that are very similar to themselves. Increasing mobility may be decreasing, not increasing, the variability of individuals within the human population!

Men, women and the future

Sociobiology and evolutionary theory tell us that there are real differences between men and women, apart from the obvious physical ones. These differences may or may not be advantageous today in evolutionary terms, but they are part of our inheritance because they gave a selective advantage to our ancestors. Women are indeed more nurturant than men, more protective, and less inclined to take risks. Men are more inclined to take risks, to show off and to be competitive. Men are more likely to be sexually promiscuous than women, and to be less forgiving than women if their partners indulge in a sexual liaison outside the pair. Some of the greatest achievements in human literature, and more than a few popular songs, are built around these themes.

This is an inevitable consequence of our mammalian system of reproduction. A man who strays from the matrimonial bed but returns home may continue to be a good provider and effective assistant in raising a family. But a woman who strays in this way may produce a child which is no relation to her partner, but which he wastes time and effort helping to rear, at the cost of an opportunity to pass on his own genes. If a woman could get away with it, though, there might well be some evolutionary advantage to her line if she found a way to mate with a powerful, successful male whose resulting progeny was then reared by her unsuspecting, lower-status husband as if it were his own. We are not suggesting that it is morally correct for people to behave like this today, or that such behaviour has any evolutionary advantage now. But because such activity did have advantages for our ancestors, it is inevitable that we carry genetic predispositions in certain directions, just as lions are genetically predisposed to kill cubs when they take over a pride. After all, 99 per cent of our genes are animal genes.

Society has developed rules, and institutions like marriage, in order to help people to overcome these genetic predispositions and to act in ways which are more appropriate today, but for which we have not yet had time to evolve the appropriate genetic predisposition. Sociobiology can make the task of living by those rules easier, by helping us to understand both why the rules are important and why people sometimes have urges to break them.

The basic high-risk, high-stakes approach of the male also explains masculine behaviour in other areas of life. Throughout their lifespan men have a higher mortality rate than women of the same age, a reflection of the greater stakes our male ancestors played for in the reproductive game. This is a direct result of the effect of testosterone on the male body, making men more aggressive and more willing to take risks. The evidence shows up clearly from studies of castrated males.

Uncastrated men seem to live life at a more hectic pace, so their bodies burn out more quickly. The average lifespan of a group of castrated males in American homes for the mentally retarded is 69.3 years, compared with 55.7 years for a carefully chosen sample of equivalent uncastrated men in the world outside. And, indeed, the reason why castration was practised for a time in American institutions of this kind was precisely because it made the inmates more docile and manageable.* You can see the same effect at work in your domestic cat. The intact tom leads a far more exciting and adventurous life than his castrated litter-mate, who sleeps slothfully by the fire. But the neutered cat lives a longer life. Testosterone is the immediate cause of typically male behaviour, but of course the human body has evolved to respond in this way to testosterone because typically male behaviour has been successful, in terms of passing on genes to succeeding generations. The shorter lifespan has been outweighed by the increased opportunities to reproduce.

Men are more likely than women to be killed in accidents – car accidents, mountain-climbing, hang-gliding, or whatever – but they are also more prone to disease. The reason is easy to see, with hindsight. In primitive society, a pregnant woman who gets sick is likely to lose the baby she is carrying, which represents a great loss of her reproductive potential, even if she soon recovers. A man who gets sick may feel dreadful for a while, but gets off his sickbed and impregnates another female the next week. And then the implication for a young infant's prospects of survival if its mother dies, cutting off its food supply, is very different from the implication if its father dies. There has been a strong selection pressure for women to be fit and healthy, but far less so for men, and in evolution things that are not selected for do not, by and large, evolve.

Extreme versions of male and female patterns of sexual behaviour are found among homosexuals. Homosexual men have tended to be promiscuous and have many partners; homosexual women tend to have few, or one partner in a long-term relationship. The origins of homosexual behaviour are far from clear, and provide a debating ground for experts, with whom we would not wish to cross swords ourselves. But there

* Martin Daly and Margo Wilson, *Sex, Evolution, and Behavior*, p. 75.

does seem little doubt that homosexuality does represent, in some sense, the extreme sexual strategies of males and females, unfettered by the need for compromise implicit in a heterosexual relationship.

So what should we do about the differences between the sexes? We are not saying that they are 'right' today, only that they were the successful strategies selected by evolutionary pressures acting on our ancestors. In modern society, relationships between men and women (or between members of the same sex) can take very many forms, and still healthy children are brought into the world and grow up to have children in their turn. We learn from sociobiology that men and women are born with different innate abilities, inclined in slightly different directions which correspond roughly to the stereotypical roles of male and female. Society, even today, acts to emphasize those differences, to make men more masculine and women more feminine. But those roles are no longer essential in society, nor are they necessarily appropriate. A woman can head a corporation, or fly an airliner (but how many do?); a man can be a househusband, or in charge of a crèche (but how many are?).

The key to a happier society, for both men and women, lies in the strength (or, rather, weakness) of the genetic predisposition towards male or female roles. In fact, the predisposition is small, as has been shown by studies of boys reared as girls after unpleasant accidents during circumcision, and of genetically female girls who have been subjected, for various reasons, to the influence of male hormone early in their lives as fetuses. The differences between the sexes are real, but small: as sociobiologist Edward O. Wilson has put it, 'at birth the twig is already bent *a little bit*' (our italics).* But there are enormous differences in the exact roles of the sexes in different human societies, and the success of the cultural influences in 'training' those boys to be 'feminine' shows that, as always seems to be the case, both nature and nurture are responsible for the adult. The sexual division of labour, says Wilson, 'is not entirely an accident of cultural evolution', but there is a clear implication

* Edward O. Wilson, *On Human Nature*, p. 132. The second quote is from the same source.

that cultural influences can work with or against the slight biological inclination of a developing human being towards masculinity or femininity.

Our society has been and still is, dominated by men, and has also acted to exaggerate sexual differences in behaviour. Even women who make it to the top of our society, such as Margaret Thatcher, have done so by being apparently more masculine than the men, almost caricaturing the supposed male attributes – in this case, the 'Iron Lady' was described at the time of the Falklands War as being 'the only real man' in the government. We believe that the lesson to be drawn from a sociobiological perspective on the origins of sex roles is that society can and should take positive steps to reduce the natural differences. Culture can go a long way towards creating a more equal society, but this will not be achieved simply by passing laws which say that men and women are equal. It has to start at the beginning, with education. Boys should be actively encouraged to develop traditionally feminine interests and skills; girls should be encouraged to take up things traditionally associated with boys. Only then will we be using our 1 per cent advantage over the apes to effectively circumvent our animal genes. But we shall have to use that 1 per cent to its utmost if we are ever to work out exactly how this should be achieved. For example, many proponents of sexual equality fought long and hard, in many countries, to ensure that boys and girls are taught together in the same classrooms. We believe that they should be taught separately, in single-sex schools, for two reasons. First, boys and girls develop, physically and mentally, at different rates; it is ludicrous to teach mixed-sex classes of children, grouped on the basis of age alone, between the ages of about 10 and 18. Secondly, largely because of the culture we are at present stuck with, girls tend to defer to boys even at an early age, and several studies have shown that girls in single-sex schools do better (given the opportunity) in traditionally male areas such as science than do girls in mixed schools, where the boys dominate such activities.

Shaw's Professor Higgins asked, 'why can't a woman be more like a man?' Our answer is that she can, and so can a man be more like a woman. The world would certainly be a happier place if both sexes were brought up to accept a more equal middle role, but it is very hard to see how such a desirable state of affairs can be achieved, starting out from our present position.

CHAPTER NINE

Putting People in Perspective

Why is the helpless, dumb blonde, as portrayed so memorably by Marilyn Monroe, a successful role for some individual members of human society? The reasons lie in our genes as much as our culture, and have to do with the size of our brains and the resulting helplessness of human babies.

Human babies are physically helpless. They survive only because their helpless appearance evokes a strong response from most adults. Even male chauvinists who profess a profound dislike of infants will coo over a newborn baby, responding to some deep-seated biological imperative, while a mother is able to identify the whimper of her own baby in a busy maternity ward, and when at home will wake at the first cry in the night, even if she was previously a heavy sleeper. It is easy to see how such responses must have evolved in tandem with the helplessness of the human baby at birth, a helplessness which is itself, remember, directly related to the evolution of the large brain that makes us so successful as adults. In primitive societies people who are repelled by helpless infants do not, for obvious reasons, leave many offspring to carry their genes forward to future generations. Some men may, but scarcely any women, before the development of a civilized society in which it is possible for child rearing to be taken over almost at birth by people who are not the biological parents of the infant. So parents who respond more lovingly to the helpless appearance of their own babies will be more effective at raising children and ensuring the spread of their own genes, including the package of interacting genes that makes them love babies. And this raises the possibility of an interesting evolutionary feedback.

The successful smile

The kind of feedback can best be understood by considering the peacock's tail. This wonderful accoutrement is used by the male bird solely as a lure to attract females and obtain a mate. The female birds are obviously attracted by the tail. We do not have to know why this should be in order to understand the feedback process, which is just as well since some of the arguments put forward to 'explain' the phenomenon are quite tortuous. Some have suggested that the appearance of the tail, the colours of its feathers making a pattern of a myriad eyes, may hypnotize the hen into submission; others argue that since such a large tail is a physical handicap to the peacock, only an otherwise very 'fit' bird could survive while carrying such a handicap, so females have been selected to mate with the males with the biggest tails. Whatever the reason, though, in the world today peahens *do* prefer to mate with peacocks with the largest, and most colourful tails. Because the peahens are, in human terms, obsessed by this characteristic, any male with a bigger tail will mate successfully and have many offspring. So genes that provide for large, colourful tails spread rapidly through the population, and the present extravagant display can evolve from a much smaller feature over many generations.

Let us return to parents and babies. Successful parents are those that respond to the helplessness of their babies with love and attention. So packages of genes that promote this parenting response are widespread in human population, almost universal. In such a situation it may actually be in the best interests of the baby to seem even more helpless than it really is, because that will evoke a stronger parenting response. The baby, in other words, may actually be psychologically manipulating its parents from the moment it is born. It has to: as Robert Trivers has pointed out, because the infant is much smaller and weaker than its mother, it cannot physically fling her to the ground and suckle when it is hungry, so psychological weapons are the ones it must use to achieve its objectives.

Of course, even newborn babies are not completely helpless. They can suck, cry, see, hear and grasp. They really do look even more helpless than they are. But the relationship is not all one-way. Evolution has provided the baby with a powerful means of rewarding its parents for all the attention it gets: the smile. Crying communicates a baby's distress, and this essential means to stimulate adults into action is present at birth. Within a very few weeks, however, the baby also develops a smile, so heart-warming that many adults (especially women) other than its parents respond to it. The smile encourages adults to interact with the baby, play with it, coo over it and tickle its toes. It also makes the adult feel good; it is a reward to the adult for bothering to pay attention to the infant. And, once again, it is easy to see how such a system has evolved through a feedback process like the one which has made the peacock's tail so big. There is no conscious thought involved, any more than a peacock thinks, 'Hmm, I'd like to grow a big tail.' It is all instinctive, coded in our genes. Back in evolutionary history, once babies started to smile (and it may have been simply a grimace connected with wind pain to start with), then adults who responded to the smile were favoured by selection because they gave their babies more attention and helped them to grow up successfully. So infants that smiled more were favoured more, and so adults who responded were further favoured, generally speaking, and so on. The result is that we all carry genes which ensure that we smile a lot as babies (whatever culture may do to change that pattern of behaviour as we get older), and we all carry genes that make us respond warmly to smiling people, especially helpless, smiling infants.

Which brings us back to Marilyn Monroe. Just as infants are physically weaker than their parents, so women are, by and large, physically weaker than men. Abhorrent though the image may be to many women today, the stereotype of a big, husky male, and a weaker female is unfortunately close to the truth, and just as babies use psychological weapons to get their way with adults, so women can, consciously or unconsciously, use psychological weapons rather than brute force to get their way with men. And what better weapon to use than the one which has proved such a success with the babies: an appearance of helplessness, combined with a fetching smile. The success of the not-so-dumb blonde in melting the heart of a husky he-man is no surprise at all to the sociobiologist; it simply reflects a distortion for adult use of the infant's prime weapons in the generation game.

Forming the bond

If anyone doubted the essential need for a newborn baby to latch onto the set of instinctive responses that makes its mother love it, a study carried out by two doctors in Cleveland in the early 1980s underlines the power of the mechanism. Marshall Klaus and John Kennell started out from an initially baffling puzzle: although improved medical care was enabling doctors to save many premature babies that would have died even a few years before, these infants were far more likely than full-term, normal infants to return to hospital as victims of baby battering or neglect. It was as if the long separation of the baby from its mother while it was in intensive care in an incubator had prevented the normal instinctive bond between mother and child from forming at birth, and by the time the babies were united with their mothers it was too late for the bond to form.

Again, we do not need to go into the means by which the bond-forming process is triggered. Perhaps it is related to hormones present in the mother's body at birth, but not a few days or weeks later. But one interesting feature noted by the Cleveland team* is that newborn babies are remarkably alert for the first forty minutes or so after birth. Their eyes are wide open, bright, and move to follow a moving object, especially a human face. A mother holding her newborn infant almost always concentrates on its eyes, and is often heard talking to it, saying something like 'Just open your eyes so I'll know you're alive', even though the baby is quite obviously alive and kicking. The mother automatically aligns the baby so that she can look into the eyes, and the bonding process seems to be based on eye contact between mother and baby immediately after birth: literally, love at first sight.

So Klaus and Kennell have pioneered a more 'natural' approach to mother–baby relationships within the hospitals where they work, an approach which is now gaining much wider acceptance. There are still problems with tiny, premature babies that must be kept in incubators if they are to survive at all. But some mothers are now able, and are encouraged, to spend time alone with their baby immediately after birth, and to spend much longer each day with the newborn infant than

* Reported in *Annual Editions: Biology*, edited by John Crane, p. 72.

has been the norm in hospitals. The first infants to benefit from this approach are now reaching school age, and they seem to show the benefits of close mother–child bonding. Continued monitoring of their progress over the past few years has shown that the so-called 'early contact' mothers have been more attentive to their children, with one result that the children have richer vocabularies, are more confident, and have been stimulated to the point of having a measurably higher IQ than their peers (a complete vindication, incidentally, of Binet's original ideas about intelligence). And, indeed, the incidence of baby battering among this group is zero.

These are all clearly good things from the point of view of the adult as well as the child. But some of the most interesting evolutionary insights into the relationship between parents and children come from studies of areas of conflict, where the needs of the child and the needs of the adult are not the same. It might seem that what is good for the child must be good for its parents, who want to ensure that it survives to carry their genes into the future. But this misses the point that parents are also able to 'invest' in other children. To the parent, one child is as good as another as a gene carrier; but each child, sharing only half its genes with a sibling (or indeed with a parent), 'cares' much more for its own good than for theirs, in evolutionary terms. And that is where things get interesting.

The battle of the generations

This is still a new and developing area of scientific research, and the detailed implications, especially for human beings, are very far from being fully worked out. But that does not make the broad outlines of the work in progress any less fascinating. It all began in 1974, when Trivers published a scientific paper on parent–offspring conflict. He has since developed the theme more fully, but the best succinct statement of what the idea is all about is in Martin Daly and Margo Wilson's *Sex, Evolution, and Behavior*. The important point is the degree of genetic relatedness between different individuals in a family group. It does not have to be a human family, although that is the sort we are especially interested in. People are animals. We have exactly the same kind of genetic material as other animals, and share up to 99 per cent of the DNA with our closest relatives. We obey the same evolutionary rules as other animals, and even the things that set us apart from our closest relatives are not all unique to our species. As we have already seen, people are in the minority in the animal world as a species where the male takes on a degree of responsibility for the offspring. But this is a respectably large minority of species, and for this reason we are also especially interested in the behaviour of birds, where there is a similar sharing of parental responsibility. But in bird or human or other species, the only way in which an individual can be successful, in terms of evolution, is to leave copies of his or her genes, and to be sure that his or her children have reproduced in their turn. Leaving aside the possibility for doubts about paternity, each parent has a 50 per cent genetic investment in each child; parent and offspring have a genetic relatedness of $\frac{1}{2}$. On average, two offspring of the same set of parents also have a relatedness of $^1/_2$. Without going into the mathematical details, we can see that in everyday terms each parent 'needs' to raise at least two offspring in order in order to ensure that all its own genes are copied, and would 'like' to raise more. Each infant, on the other hand, only really cares about its own survival, although it will be happy to see siblings also doing well, but provided their success is not at its own cost. In other words, the strategy by which a parent can maximize its own reproductive success is not necessarily compatible with the strategy by which the child maximizes its own reproductive success. In particular, parents will try to produce more offspring when existing children prefer, in evolutionary terms, to see more attention given to themselves than to new siblings.

At some point the growing infant becomes able to stand on its own two feet, or fly on its own two wings, and make a living without further parental help. The parent's interests are best served by getting the offspring off her hands (usually, it is the mother that is primarily concerned here) as soon as possible, and getting on with the job of raising another infant. The offsprings's interests are best served by getting the mother – and father if possible – to continue to provide help, or food, for as long as possible. But the offspring do share part of the mother's investment in their siblings, so there comes a point when, *even in terms of the best strategy for their own genes,* it makes better evolutionary sense for them to cut loose from the proverbial apron strings and make their own way in the world, so that the parents can raise more offspring unhindered. Because of the simple relatedness of $\frac{1}{2}$, each infant should be inclined to seek aid from the parent as long as the cost to the parent, in reproductive terms, is less than half the benefit to the infant. But once the cost to the mother is more than twice the benefit to the offspring, the infant should be willing to leave her to raise another sibling. To the mother, however, the time to start concentrating on the next infant is when the benefit of doing so is equal to the cost of continuing to help an older offspring. And therein lies the conflict.

None of this, of course, is calculated, neither in people nor in birds. It is all instinctive behaviour, coded in groups of genes and modulated by environmental circumstances that together make animals respond in a certain way to certain stimuli. But over many generations, in both birds and people, genes that code for responses which make infants seek independence at a certain age will spread more widely than those which make infants seek independence too soon or too late; similarly, genes that make adults force independence upon their offspring too soon, say, will not spread as widely as those that code for behaviour that makes infants independent a little later in their development.

This simple language sometimes causes confusion by making it seem as if the genes are in some intelligent way directing operations. Perhaps we should stress that complex patterns of behaviour are unlikely to be caused by a single gene or simple group of genes, but that nevertheless genotypes which provide animals with certain responses to different situations will be selected by evolution and spread. But this does not mean that there is 'a gene' for leaving home at 18.

As Daly and Wilson put it:

> What this theory suggests is that there is a stage of conflict that begins at the point where the mother's investment costs her more than it gains her and ends at the point where it costs the young more than it gains them (by virtue of costing the mother twice what it

> gains them) . . . there is indeed a stage of conflict very like this, and it has a name – weaning.*

Weaning, in this sense, can be taken both literally and in a broader sense, applying to the very long period of human childhood and the comparably long period during which the adolescent readies himself or herself to leave home. In more general terms, Trivers's theory suggests that any child will always want to get more from its parents than its parents are willing to give.

Of course, we have oversimplified. The situation in the real world is far less clear cut than this black-and-white image, and there are biologists who argue that Trivers is completely wrong. Their counter-argument is that genes that express themselves in children in such a way that the child gets more than its fair share of parental investment cannot succeed in the long term, because when the body they inhabit matures and produces children those children will, in turn, extract more parental investment than is good for the parents. Such a mutant form might be imagined as producing families (bird or human) in which a very few children are raised in great comfort, while other families are busy raising many offspring in less comfort. And in that picture, the genes that enable the offspring to cheat its parent will soon die out. Such a viewpoint sees the parents as having ultimate control of the situation, since they can, if it comes to the crunch, abandon an over-demanding offspring and start again. That, too, is an extreme black-and-white view, which appears to be diametrically opposed to Trivers's argument at its simplest. The truth undoubtedly lies somewhere in between, and finding out just where in the grey area human patterns of behaviour are established is one of the intriguing lines of enquiry now being followed.† We can, however, see clear

* Martin Daly and Margo Wilson, *Sex, Evolution, and Behavior*, p. 157.

† We find Dawkins's argument, put forward in *The Selfish Gene*, persuasive. He points out that, although siblings may be competing for food, there are complications to their rivalry. An older sibling, for example, may have less urgent need of the food than a younger one, and in some ways may be acting more like a parent by ensuring that the younger sibling does well. If the net cost to brothers and sisters of the eldest grabbing the food is more than twice the benefit it would reap by having the food, then the selfish-gene theory says that the eldest should be expected to let them have the food, since there is a 50–50 chance that any of its genes,

evidence that there is a genetic tendency for self-restraint in the use of resources within a normal family unit, but also that parents are not in complete command, by looking at the example of the cuckoo.

Young birds in a typical nest may seem to compete aggressively with their brothers and sisters, with their loud cries and broadly gaping mouths as they beg for food, but the cuckoo takes this to extremes. The female cuckoo lays her egg in a carefully chosen nest built by another species of bird, and containing the other bird's eggs. So a young cuckoo hatches from its egg in a nest alongside the eggs laid by a bird of another species. It has no interest at all in seeing these foster-siblings survive, nor has it any direct genetic interest in whether or not its foster-parents survive or reproduce. It removes its potential rivals for food by pushing the other eggs out of the nest, and it makes extraordinary demands on the 'parents', working them to exhaustion as they respond blindly to the 'psychological' pressures brought to play upon them: the presence of the great chick in the nest, its loud cries and enormous gape. The example of the cuckoo serves to remind us how genuine offspring are much less selfish than they might be in their demands for parental care, but it also shows us that the parents are conditioned by their evolution to the point where they cannot simply reject the nestling and fly off to raise another brood.

Old fogies and young tearaways

Trivers himself has described his observation of parent–offspring conflict connected with weaning in species as diverse, and geographically far apart, as pigeons living wild in Massachusetts and monkeys in East Africa and India.* Young pigeon chicks are fed carefully by both parents, who bring food to the nest and encourage the chick to feed by stroking its neck with their bill. But when the chicks are almost fully fledged it is

including the gene for such 'altrusim', resides in their bodies as well. This is the kind of 'grey' argument that tells us more about what is going on than either black or white argument.

* Robert Trivers, *Social Evolution*; see e.g. Chaper 12.

they who pounce upon the parents as they return to the roost, demanding food and crowding the parent into a corner. Indeed, says Trivers, he often saw these city-dwelling pigeons ignore their almost fledged offspring, and their own nest, to fly instead to some different roosting ledge on a nearby building in order to escape this harassment.

Similar incidents happen with langur monkeys and baboons. It may take several weeks for a mother to break her infant of the habit of demanding milk, or a ride on her back. Outline studies of these and other species show the same broad pattern of behaviour predicted by the theory based on kinship, and where more detailed studies can be made they often bear out the theory very accurately. In most cases, it seems that the result of parent–offspring conflict is that the young get more from their parents than the parents would ideally 'like' to give, but that this is still less than the offspring would 'like' to receive if they had things all their own way. The good old compromise rules.

Trivers also looks at the psychological weapons that the infant has in its armoury, one of which seems to be the temper tantrum, no less a feature of chimpanzees (and other primates) than it is of people. A young chimp or human child will scream with rage, fling its body to the ground, and bash itself against hard objects. This is all very alarming for a parent, because there seems to be – and may even genuinely be – a risk of the infant being injured. Parents who fail to respond to a real risk to their offspring will suffer in the evolutionary stakes, so the temper tantrum may trigger the protective parenting that would be a correct response to a real risk to the infant. In this case the strategy may be as much calculated as it is instinctive. Chimpanzees as well as human children have sometimes been observed, before or in the middle of a tantrum, having a quick look round to make sure their antics will be observed. Such tantrums should be most common precisely during the period of conflict, what we have called 'weaning' in the broader sense, when the infant still wants parental attention but the parent would be better off investing effort in younger infants. That is exactly what is seen among chimpanzees, and will certainly have a familiar ring about it for human parents.

All this has very important implications for the psychological and social sciences, implications which have not, as yet, begun to work their way into the teaching of those disciplines. Conflict between parents and offspring is now seen as inevitable, but for very different reasons than those enshrined in, say, Freudian psychology. Deceit, intentional or instinctive, is also an inevitable part of the generation game, with children pretending to need more from their parents than the parents are willing to give freely. And there is an inbuilt tendency for individuals to act out the roles of their own parents when they become parents themselves, even though they strenuously acted against their own parents when young! Caught in an inevitable evolutionary spiral, today's young tearaways become tomorrow's old fogies, because the best evolutionary strategy for being a successful adolescent is not the same as the best strategy for being a successful parent.

The parenting puzzle

Most of the parent–offspring conflicts arise in present-day human society when the offspring reach adolescence. The teenage years correspond very closely to the broad definition of 'weaning' as the time when the offspring wants the parents to provide for it, but the parents would rather make other use of their resources. In our society, of course, this does not necessarily mean investment in more offspring, although that is still an element. Many people today, child-rearing off their hands and still – unlike our counterparts of just a few centuries ago – with many active years ahead of them, would like to take the opportunity to live it up a little and enjoy themselves.

This is a quite new phenomenon, in evolutionary terms, since even for human beings the usual fate of females, in particular, has been to keep reproducing as long as possible and then drop dead from exhaustion. The phenomenon of an extended and healthy middle and old age is still too recent for there to be any clear indications of how our genetic imperatives, selected for quite different ends, will allow or encourage us to respond to this new-found freedom in middle life. But it is certainly interesting that so many empty-nesters, given the opportunity, behave in ways reminiscent of young, newly mature people:

taking holidays, taking up new sports, living more dangerously, even finding a new sexual partner. It is as if the pattern of behaviour that is triggered by the emergence of a juvenile from the cocoon of family life into independent adulthood can also be triggered by the emergence of an adult individual, or a pair of parents, from the responsibility of family life once the children are raised. And this is just the kind of thing we expect to find in human behaviour: underlying animal patterns, corresponding to our 99 per cent inheritance, deflected into new forms by the results of our 1 per cent advantage, in this case by the medical skills that ensure our survival to a ripe old age, combined with the related skills that make it unnecessary for a woman to breed herself to death. Our chance of a 'second adolescence', with all the benefits and none of the drawbacks (not even a nagging parent!), is one of the greatest personal advantages bestowed by that 1 per cent of DNA that makes us uniquely human. Studies of how people behave in their forties and beyond could clearly be fruitful for sociobiologists. But what of the role of the growing teenager in the parent–offspring conflict?

A lot of what we have been saying in this chapter is encapsulated in a paragraph from an article by Glenys Roberts in the London *Evening Standard* on 15 August 1985. Writing on the 'Woman' page, and bemoaning the state of a world in which teenagers 'are dependent long after they are physically grown', she clearly did not appreciate that this is not a problem of modern society, but has been the pattern down the ages, in other species as well as our own. But she summed up in one sentence the view of the parent who has made enough investment (from the parent's point of view) and is waiting for the fledgling to leave the nest, even though the fledging feels that more parental investment is appropriate: 'Faced with a strapping apparition who cannot even be bothered to wash up a cereal bowl, ask yourself who it is that is being exploited.' *That* is what Trivers's theory of parent–offspring conflict is all about!

David Barash also provides some examples of behaviour patterns in our close relatives that seem to echo familiar human themes. Recall that in the wild the crucial thing from the point of view of both parent and offspring is not the chance to buy a sports car with the money you used to spend on the kids, or to use the money to take a world cruise, but simply reproduction. According to Trivers, there inevitably comes a time when mother is ready to have another baby, but her existing offspring would do better, genetically speaking, if she concentrated on them. And what we find, says Barash, is that 'immature chimpanzees frequently harass copulating pairs'.* What is more, 'immature animals responded with particular intensity to their mothers' first renewal of sexual behavior following their own birth and suckling'.

Once again, this fits in with the theory, and once again we all know all similar examples in human society and in human literature, which of course reflects human life. The issue is more clearly focused when we consider the problems that many divorced or single mothers have in getting their young children to accept a new man into the family. In genetic terms, an infant should be even less willing to accept the prospect of the arrival of new babies that have a different father, because they will share only a quarter of the existing infant's genes. Of course, there are also sociological and psychological reasons for a young child to reject its mother's new lover. But counsellors trying to advise a mother in this predicament generally start out from the assumption that the child's jealousy is solely due to the attention that is diverted from it to the mother's new partner. Triver's theory suggests that there is also an in-built 'jealousy' related to the possibility that the mother will produce new babies and lavish attention to these half-siblings while ignoring her existing child.

Again, we must stress that none of this is reasoned out by the child or by the genes. The full scenario would be that children who act to prevent their mothers from finding new sexual partners have in the past received more attention from their mothers; their genes, including the ones that make them behave unpleasantly to prospective stepfather, have therefore spread (the argument runs the same, of course, when applied to stepmother as well). Sociobiologists are always urged to take the psychological and social factors into account, and rightly

* David Barash, *Sociobiology and Behavior*, p. 332.

so. But we would urge the psychologists and sociologists to take more account of the biological roots of some of the problems they encounter, the better to resolve the conflicts by using the advantages – such as reasoning ability – that we have over those animals that are more blindly directed by their genetic inheritance.

To be fair, this is beginning to happen. Daly and Wilson have looked at the Canadian statistics on children that are physically assaulted – battered – by their parents, and find a high proportion of cases in which the 'parent' responsible is a stepfather. This is understandable in sociobiological terms, because the stepfather has no genetic stake in the child and benefits no more (in those terms) from the child's presence in the family than does a cuckoo from the presence of other eggs in the nest. Harsh though it may seem, there can be no doubt that selection has operated, if only to a mild extent in the human line, to favour individuals that are hostile towards prospective step-parents, and simultaneously to favour individuals that are hostile to their stepchildren. And these in-built hostilities still exist, even though in modern society the stepchild may benefit enormously from the arrival on the scene of a new partner for the existing parent, who may be well able to cope with raising the existing child *and* any others that the couple may choose to have. This is a key new factor in modern human society, being the element of choice that we have in such matters but which other animals do not.

None of this means that baby battering by stepfathers is either inevitable or natural, let alone desirable. Instead, it offers ways to minimize the risk of this happening. If nothing else, then, armed with this kind of insight, hard-pressed social workers should find it more easy to identify families and individuals at risk, and to provide appropriate counselling advice before the instinctive hostility between step-parent and stepchild breaks out into damaging conflict.

There are so many examples of the kind of behaviour that Trivers's theory predicts that they would fill a book in their own right. Looking back, again, at our closest relatives, chimpanzee mothers that give birth relatively late in life are invariably more inclined to tolerate the demands of their new infant, which weans late and has the benefit of being carried around for longer than is usual. The underlying reasons for the evolution of this pattern of behaviour are now clear: such a late baby may be the mother's last chance to reproduce, so her best course of action is to secure the existing investment rather than push the infant out into the world and gamble on the slight prospect of getting pregnant again. And although we now have choice and control over whether or not we reproduce again (or at all), human mothers show the same affection and extra care for the child that comes late in life. Who does not know a family with an indulged youngest son or daughter? Older parents and their children suffer less from parent–offspring conflict, not just because the parents are older and wiser, but also because they are more willing to let the children have their way.

But when there is conflict, it will follow the usual pattern. Parents in general will want their children to be less selfish, because to the parent it is a good thing for all the children to be treated equally. Each child will, however, tend to be more selfish (and spiteful) than the parents think desirable, because he or she shares only half of his or her genes with even full siblings. This dichotomy actually extends to more distant relations as well: a child's cousins share only one-eighth of his or her genes, but the child's parents share one-quarter of the genes of what is, to them, a niece or nephew. So it is very broadly true that parents will expect, or want, their children to behave less selfishly to the children they are most likely to come into contact with.

This example is rather nice, since it also shows how biology is subverted by culture. During the many generations over which this process was evolving, the children that a child was most likely to come into contact with were indeed siblings, half-siblings, cousins and other relations. So the instinct that has evolved is for parents to encourage children to be nice to their playmates. In the past, this would have helped the spread of copies of the parents' genes. Today, those playmates are unlikely to be relations, but the instinct is still there, and we still exhort our offspring to be nice to other children, to share their toys, and so on. We tell ourselves that this is just 'the right way' to do things, and there may be an element of reciprocal altruism involved. But at heart this is rationalization of our automatic behaviour. We are following an instinct that evolved under quite different circumstances.

Parents also want their children to be successful. Children represent the investment of the parents' genetic capital in the future, and as the parents get older it gets less and less likely that they will get more chances to reproduce. Teenagers, on the other hand, see their lives stretching out ahead, and want to have a good time while they are young; there will be plenty of time to settle down and raise a family later. So parents inevitably cajole or exhort their children to study hard, to keep out of trouble, not to drink or take drugs, to be nice to other people, and so on. These are all courses of action that would actually improve the immediate prospects of the parents' genes being 'looked after' and passed on; in the jargon, they would 'maximize the parents' genetic fitness'. We shall leave you to draw your own conclusions about why people behave as they do in your own family. But there is an important, and encouraging, aspect to all of this.

Some people worry about the implications of sociobiology and the concept of the selfish gene, because these ideas revolve around the evolutionary fitness of individuals, and the selection of individuals in competition with one another. As we have seen, this inevitably provides scope for conflict, and the worriers fear that this not only 'explains' human aggression, up to and including war, but makes it inevitable. But we have seen that a tendency for parents to exhort children to be nice to one another, which has evolved for sound 'selfish' reasons, has been channelled by society into a more generally altruistic form, one which our reason tells us must be a good thing. We do not have to go through the sometimes painful process – fatal for many individuals – of evolving a new set of responses to adapt to a changing world, because we can see for ourselves that the right course of action (that is, the one that will bring most benefit to ourselves) is almost always (today) one of cooperation for mutual benefit: with another individual, or group of individuals, or even other nations. The selfishness of our genes may indeed have been partly responsible for bringing us to the edge of nuclear war, because our instinctive responses to conflict evolved in circumstances where a stone axe was the ultimate deterrent. But there is every prospect of using the advantage our intelligence and reasoning ability gives us, especially now that the Cold War has thawed, to avoid the ultimate conflict. As we shall see, the threat might all have been a mistake anyway.

Aggression is one of the hardest subjects to tackle in any discussion of sociobiology because it is widely misunderstood. The image of 'nature red in tooth and claw' still seems to go hand in hand, in the minds of all too many people, with the idea of evolution by natural selection and survival of the fittest. But, as we have explained, most of the competition in nature involves breeding strategies, not aggression. Certainly there are aggressive species, and ruthless 'warfare' is waged at some levels in the 'struggle for survival'. But these are not the levels at which large mammals like ourselves operate. The most vicious 'aggression' seems to be the prerogative of rather smaller creatures – wasps which lay their eggs in the living flesh of larvae from other species, so that their young will grow up in a ready-made, living food store, or the bugs that invade our own bodies and cause disease. Even the bloodiest of mammals seem noble by comparison, as we recognize in the names we apply to our fellow human beings.

Who, after all, would not be happy to be described as 'lion-hearted', 'cat-footed' or 'strong as a bear'? Predators – animals that eat other animals smaller and weaker than themselves – are generally respected and held in esteem or awe. But parasites – animals that eat other animals larger and more powerful than themselves – are despised. The very name 'parasite' is a term of abuse, while calling a man a louse is asking for trouble. Yet a louse is no less intrinsically courageous than a lion, in its way. Our choice of epithets tells us as much about ourselves as about the animals whose names we use as labels in this way, and shows that the kind of aggression practised by creatures like lions and bears is something that seems normal in human terms. Clearly, people are in some sense aggressive. Warfare and sporting substitutes for warfare are part of our culture, and in order for them to be so there must be something in our genetic inheritance that predisposes us to fit in with this kind of cultural pattern. But just how aggressive are bears and lions?

Cowardly lions

Ecologist Paul Colinvaux has given an insight into the true extent of red teeth and claws in nature in this book *Why Big Fierce Animals are Rare*. Animals, he tells us, come in distinct sizes, with gaps in between. A fox is ten times bigger than the songbird on which it preys; the bird is ten times bigger than the insects it eats; and one of those large insects will be ten times bigger than the mites which are its own prey. This scaling is understood in terms of the need of the predator to be fast enough to catch its prey, and big enough to swallow it, more or less, at a gulp. Of course, this is an oversimplifaction. You will have to read Colinvaux's book if you want the subtleties, and an explanation of the (few) exceptions to the rule. Bigger animals in such a chain are disproportionately less common than smaller animals. You might think at first that there would be one-tenth as many foxes, say, as there are songbirds; but the larger predators are actually much rarer than that. The reason seems to be their need to be able to run fast (or fly or swim fast) in order to catch their food. Being a predator is an energetic lifestyle that uses up a lot of calories, so each predator needs a large territory over which it can hunt for prey.

All the energy in living things on Earth, except for a few strange creatures that live in the murky ocean depths by hot volcanic vents, comes ultimately from the Sun. Plants convert solar energy into what animals regard as food. Many animals eat plants to get their energy, and other animals eat the plant-eaters. But at each stage up the chain energy is lost. No conversion process is 100 per cent efficient (quite apart from the energy used up by creatures at each stage in living their own lives), and by the time plant energy has been converted into animal bodies that have in turn been used as food by other animals, there has been a great deal of wastage. So there is a strict limit on how big a predator can be and still be able to eat enough to stay alive, and that limit is at about the size of a tiger, a bear or a shark. Plant-eaters, such as elephants, can grow much bigger than this because they do not have to run fast to catch their prey – it just sits there, helplessly waiting to be eaten – and also because they are dealing with the basic food source at the bottom of the chain, the next best thing to solar energy itself. And whales, the largest mammals (and largest animals, today) of all, cheat by cutting out the middle-men. The great baleen whales, as Colinvaux points out, do not eat prey one-tenth their own size. Instead, they use their mouths as sieves to strain much smaller creatures, the shrimp-like krill, out of the ocean waters by the tonne. So they cut out the energy losses that would have been entailed if the krill had been eaten by fish, which were eaten by bigger fish, which were eaten by whales.

Colinvaux even has an ingenious explanation for the success of the dinosaur *Tyrannosaurus rex*, which many a child's book of dinosaurs will tell you was a big, fierce flesh-eater. Big, yes; flesh eater, yes; but fierce – no, according to Colinvaux. It seems that the latest thoughts on the *T. rex* skeleton are that it was a rather slow-moving creature, not at all fleet of foot, and that it probably therefore fed on the dead carcasses of plant-eaters, perhaps finishing off the sick or dying, but not engaging them in fierce combat at all.

This gives an insight into the lifestyle of most of the 'fierce carnivores' that are around today. Lions, tigers, and the rest generally do pick off the weakest prey animals when they go in search of food. The very young, the very old or the sick get culled from herds in this way, but only on the rarest occasions is a healthy, mature adult pulled down. Predation is really only a glorified form of scavenging. Studies of wolf packs hunting deer or moose show why: if a healthy adult male stands and fights he will be overcome by the pack, but not before he inflicts injuries on some of his attackers. The injured wolves will be unable to hunt with the pack, and will die. So, even though the pack can together bring down a healthy adult male, evolution has selected favourably those wolves which choose not to pick on healthy adult males, and has selected against those wolves that do. Lions too are evolved to be cowardly, and never to pick fights with big, strong antelope! There are natural in-built limits to aggression in the fierce animals that we regard instinctively as noble role models, and there are similar natural in-built limits to agression in human beings. The problems we face today, of course, revolve around the fact that those instincts are based on millions of years of evolution during which people did not have access to technological weaponry, and so we have to use our intelligence instead of our instincts to decide how much aggression is in fact 'worth while' today.

Tribes and nations

Human aggression exists because it has been selected for in our tribal past. As with other species, we were – before the invention of civilization – predisposed to act aggressively when the likely costs were low and the likely rewards were high, or as a last-ditch, all-or-nothing response to a perceived threat. When there are only limited resources to be had, aggressive behaviour comes to the fore and is more likely to be directed towards strangers, according to the opposite of the arguments about kin selection which explain why we have evolved altruistic behaviour towards our nearest and dearest. If it pays to help people you are probably related to, then it also pays to damage the prospects of those you are not related to.

But people can hardly be said to have done much damage to each other's prospects at all, before the advent of modern technological warfare, compared even with the behaviour of the cowardly lion. Edward O. Wilson stresses the point when he says that those who depict humankind as blood-thirsty in the extreme are simply mistaken. Many studies show that species such as lions, hyenas and some kinds of monkey go in for fights to the death, infanticide and cannibalism to a far greater extent than found in any human society. 'If hamadryas baboons had nuclear weapons,' says Wilson, 'they would destroy the world in a week.'* This rather suggests that, since we have lived with nuclear weapons for five decades, we are models of restraint and caution, by some animal standards! Wilson sums up by saying that the evolution of aggression has left us with a genetic predisposition towards learning some form of cultural aggression, moulded by the prevailing environmental conditions and by the history of the particular group an individual belongs to, which biases it towards one cultural option rather than another. Within those constraints people have fought in circumstances where fighting may increase their

* Edward O. Wilson, *On Human Nature*, p. 104.

own genetic success, or that of close relations. The cry 'for King and country' is precisely explained in sociobiological terms.

Investigations of non-technological societies have shown that human agression typically results in only minor amounts of fighting, only occasional injury and even rarer death. Individuals who do well in battle (if these skirmishes really qualify as 'battles') certainly do well, in evolutionary terms, by gaining status, wives and property. Those who do badly, however, are scarcely any worse off than they were before. So the selection pressures which maintain this modest amount of aggression are clear, and just as in other species (male red deer competing for mates, or whatever it might be) threats and bluff are all-important, while there are clear signals of submission which are used by losers to break off hostilities before irreparable harm. The archetypal society of the North American Indians is as good an example of this as any – before the arrival of white Europeans, 'warfare' among the tribes was so ritualized that in many cases no attempt was made to kill opponents, but great prestige was gained by a warrior who 'counted coup' by touching his opponents in the thick of battle. Such a chivalrous attitude did not stand the natives in good stead against the invaders.

But why did the Europeans invade North America at all? What are the pressures that can cause human aggression to spill over and to override the checks and balances provided by evolution? The anwer seems to lie in a combination of technology – shooting a man at long range with a rifle is sufficiently remote that any instinctive compassion cannot take control – and population growth. In his book *The Fates of Nations*, Paul Colinvaux has worked out a detailed, and controversial, theory along these lines which purports to explain most of the history of civilization in ecological terms. His argument is not universally accepted, but certainly provides some food for thought.

The argument builds from the evidence that people are unlike other animals in being able to calculate in advance how many infants they are likely to be able to raise to maturity. But, like all other species of Earth, we have an in-built drive to raise as many offspring as possible. Any species that did not carry this biological imperative in its genes has long since lost out in the evolutionary stakes. Every other species occupies a definite place in the ecological web, a niche. There is a niche for foxes, a niche for elephants, a niche for songbirds, and so on. Each niche has room for a certain number of inhabitants, and if more are produced, for whatever reason, then they will die. A hundred million years ago there were much the same sort of niches available to life on Earth as there are today, but many were filled by dinosaurs. There were dinosaur equivalents of cattle and birds, foxes and sheep. It was only after the dinosaurs died out, as a result of environmental changes, that the way was opened for new species to move in and take over these empty niches, and the ones that did so successfully, radiating out in many different forms from a stock of small species that had almost literally lived under the feet of the dinosaurs, were the mammals. The niches were still much the same, so the mammals that evolved and adapted to fill them became very similar in their lifestyles to the dinosaurs that had preceded them.

But people are different. Perhaps our greatest advantage over other species is our versatility, our ability to adapt to different niches, or even to invent new niches for ourselves. People are unspecialists. Colinvaux makes the point by drawing an analogy with professions. The 'niche' of aeronautical engineer has a certain number of places in it, and the number of places depends on the size of the aircraft industry. If schools and colleges decide to train more people to be aeronautical engineers, it does not follow that they will find jobs. The surplus will have to retrain in some other field and find work elsewhere. People, says Colinvaux, have always adjusted their numbers to match the resources available: either by infanticide or by social customs that limited sexual activity, our ancestors made sure that they raised the maximum number of children that could be fed, but did not attempt to raise too many. That would have been counterproductive, as resources would have been spread too thinly for any to survive. When first agriculture and then the various technological revolutions came along, the immediate effect was to make people rich, with more resources available for the old, small population. But the second result, he says, was that people instinctively adjusted their 'breeding strategy' to produce more adult heirs to take advantage of the new opportunities. So populations boomed, while at the same time expectations had been raised. And when the new resources, or technologies, came under pressure from the rising population, it became natural for our ancestors to turn to new lands, as the Europeans did when they moved to North America, in order to maintain the higher standard of living for as many people as possible and for as long as possible.

This is a very interesting thesis. It gives us a new view of historical events such as the rise and fall of the Roman Rmpire. Also, perhaps, it tells us something about the way people behave in modern societies. Poor people, says Colinvaux, have much lower expectations than rich people, and it takes less in the way of resources to maintain a person in poverty than in affluence. So, in modern society, even with the availability of contraceptives and abortion, and in spite of anything that they may say (or even believe) to the contrary, poor people will instinctively raise large families. This is a 'good thing' in terms of success of their genes, since they will leave many offspring (even if those offspring live in poverty) to have children in their turn. Rich people, however, have to make a bigger investment in their children in order to be sure that their offspring will be able to maintain their lifestyle. Many years of education make raising each child a long process, and the cost of the process restricts the resources available for other children. So the 'correct' evolutionary strategy for the affluent is to have a few children that are each set carefully on the road to success. And this, says Colinvaux, is why the wealthy have smaller families than the poor.

As populations rise, resources are spread more thinly and society becomes more restrictive, hedging people around with rules and regulations in order to maintain a reasonable standard for all. Writing years before the end of the Cold War, and perhaps being deliberately provocative, Colinvaux suggested that the Soviet Union, with an educated but small population and a very large landmass rich in mineral resources, 'ought' to be a haven of liberty and freedom, and probably would be before the middle of the twenty-first century, while the USA, having already used up resources such as oil and with a rising population, could follow the path towards bureaucratic totalitarianism. With the collapse of the Soviet Empire, half of his prediction already seems well on the way to becoming fact.

But there ought never to have been any reason for either superpower, even when there were two superpowers, to attack the other, because the logic of nuclear war dictates that there could be no winner, and superior logic is one of the most important features of our 1 per cent advantage over the apes. (It is 'superior' only in the sense that we can work out longer chains of cause and effect than our hairy ape cousins; yet again, we see that what separates us from the other apes is only a matter of degree. In this case, the computers in our heads are bigger than theirs, and can handle more complex chains of reasoning, but seem to run on the same basic rules.)

'The worst prospect we have to face', says Colinvaux, 'is that the freest of us will lose our liberty from a remorseless and gentle jostling of crowds of people.' But this need not happen, because

> we are human, understandable, and very different from other animals. We breed in human and controllable ways. We can change our life styles to let women do more useful things than raise surplus children. Because we can work out what is happening to us, we need fear neither our future nor our fate.*

But how, then, could we possibly have been so stupid as to get ourselves into a nuclear arms race, threatening a war which nobody could win? As we said earlier, it seems that it was all a mistake, the result of the workings of an evolutionary strategy that may be stable in the mathematical sense, but which has been distorted by the unusual circumstances confronting humankind in recent decades.

The Prisoner's Dilemma

We learned about the games-theory approach to an understanding of the nuclear-arms race through the work of P.G. Bennett and M.R. Dando, a contact stemming from a spell one of us spent working at the Science Policy Research Unit at the University of Sussex. Games theory, of course, grew up in the

* Paul Colinvaux, *Fates of Nations*, p. 223.

context of modern warfare, and simulations of modern warfare and political strategies, and later moved sideways into the study of evolution, where it has developed in parallel for more than twenty years. The more strictly biological ESS approach has a great deal to tell us about the way individuals should behave towards one another, and this, as we shall see, is relevant to the way nations behave towards one another. But let us start with the aspect of political games theory that we learned first, a version of the problem that is known as the Prisoner's Dilemma, in the context of present-day international politics.

One of the most important lessons of games theory is that, even if everyone is agreed on the most desirable course of action – for example, nuclear disarmament – a collection of individuals or nations acting independently of each other may still find it difficult to achieve the most desirable aims. As we saw in the Hawks *v.* Doves scenario, the evolutionary stable strategy need not be the one that would bring the most benefit to the most individuals. The catch is that the only way all individuals could benefit from the best possible strategy is for all to agree in advance to pursue that strategy, and to stick by the agreement. The Prisoner's Dilemma illustrates the problem.

The basic scenario can be found in all the standard texts (here we follow the version in *Game Theory and Politics* by Steven Brams). Imagine two criminals, partners in crime, who are arrested and placed in separate cells, with no means of communicating with each other. The District Attorney believes the prisoners to be guilty of a serious crime, but has no proof that will stand up in court. He needs a confession, and attempts to gain one by telling each prisoner in turn that he will offer a deal. If one suspect confesses and implicate the other, who does not confess, the confessor will go free as a reward for cooperation, while the other gets sent down for the maximum sentence, ten years. If both confess then, since the DA can hardly set them both free, each will get a lighter sentence, maybe by three years. And if neither confesses, both the prisoners and the DA know that all he can nail them with is a lesser crime for which the maximum penalty is a year in prison.

The actual numbers are not important, but they serve to illustrate the dilemma. If each prisoner can trust the other not to confess, then the best overall deal is for the two of them to remain silent. But if one prisoner suspects that the other might rat on him, it is better to confess, even though if both confess that will result in a bigger jail sentence than if neither confesses. Confession is the strategy which *minimizes* the *maximum* jail sentence the prisoner can receive, and this 'minimax' strategy is the best one, even though it ensures that the prisoner cannot receive the lowest possible sentence allowed by the game, one year for the lesser offence.

In this simple example, there is an obvious resolution of the dilemma: the prisoners can agree in advance never to confess under any circumstances. But who will hold them to the pact?* If one suspects that the other is tempted by the offer of freedom, the dilemma is back in full force. Bennett and Dando, among others, have explained the arms race in terms of a modified Prisoner's Dilemma scenario. Each side might genuinely wish to disarm, but dare not do so for fear of the consequent aggression by the other side. Even with genuinely peace-loving 'players' in the game, the strategy that minimizes the worst thing that can happen to you (the minimax strategy) may still be to arm to the teeth and be ready to ward off aggression. And this is why.

The situation is complicated because there may be a difference between what each nation wishes and what the opposing nation *perceives* as its wishes. This brings us into the arena of hypergame theory, where the rules are different because each player sees the game differently. Genuinely peace-loving 'players' who understand each other would have no difficulty reaching a stable conclusion for the game in which both have disarmed. But if each suspects the other of evil intent, then fear of being tricked into a position of inferiority maintains the arms race. Hypergame theory can put all this on a mathematical basis, in a rather exact analogy with the classic Prisoner's Dilemma 'game'.†

So the recent perilous state of the world may have resulted

* The 'no confession' rule is, allegedly, one enforced by the Mafia; in that case, the dilemma is removed because each prisoner knows that if the other confesses and is released, a worse punishment will be enforced outside.

† You can find details of this in *Nuclear Deterrence: Implications and Policy Options for the 1980s*, edited by B. Newmand and M. Dando.

not from aggressive intent, nor from stupidity on the part of our leaders, but from a misunderstanding that can now be explained in a scientific fashion. The question of what price we are prepared to pay to deter the aggressor was the overriding one in superpower politics, even though neither side need necessarily actually be an aggressor, and the situation was more like what Bennett and Dando graphically call 'mutual paranoia'. Understanding the situation is at least halfway to getting rid of the misconceptions and finding out if the other guy is really serious about wanting to disarm. The mathematical analysis suggested that each side should at the very least respond to overtures made by the other, if only to see how far that gets us down the road to peace. The way in which the whole military confrontation between East and West collapsed as soon as the US did respond in this way to Mikhail Gorbachev's offer of disarmament shows how accurately the games theorists had summed up the situation, and suggests that forty years of Cold War was, indeed, all a misunderstanding that might have been avoided if the idea had been taken seriously before. And this common-sense conclusion is very much borne out by the application of ESS theory in general, and Prisoner's Dilemma in particular, to individual evolution.

The safety of tit-for-tat

In *Social Evolution*, Robert Trivers draws an analogy between the games theory of the Prisoner's Dilemma and reciprocal altruism. Altruism is, after all, the mirror image of aggression, so it is no surprise to find the same mathematical rules effective as an aid to understanding both. In this case, the dilemma is whether or not to cooperate with another individual. If I offer him help, and he takes it but does not help me, I lose out; but if we help each other, we both gain. What is the stable strategy in such a situation? Is it best to cooperate, or to cheat by pretending to cooperate but only taking help, or what? Should my course of action always be the same, or should I modify it in the light of the response I get from other people? Robert Axelrod and William Hamilton investigated the possibilites by using computer models of this variation of the dilemma. In much the same way that computer chess programs are set up to compete against one another in tournaments, Axelrod and Hamilton took 14 different strategy models, plus some that 'played' at random, and ran them through 200 cycles of the game in a computer. It turned out that the strategy which scored the best overall was the simplest, which is called tit-for-tat. It has only two rules: on the first move of the game, cooperate; afterwards, do exactly what your opponent did last time. Tit-for-tat, says Trivers, is 'a strategy of cooperation based on reciprocity'.

In a second round of competition there were 62 entries, most of them designed to meet the challenge of beating tit-for-tat, and a total of 3 million choices run in the computer came up with the same result, confirming the superiority of tit-for-tat in an environment where strategies compete against one another. Tit-for-tat wins hands down, especially when refined so that it will 'forgive' just one unfriendly act, turning the other cheek and offering cooperation once more before resorting to copying the other player's approach. In the animal world, this is equivalent to saying that animals following the tit-for-tat strategy would have an advantage sufficient to wipe the evolutionary floor with all others. It is an ESS with a vengeance, and no competing strategy can displace it once it has become established.

This has profound implications for everyday life, as well as for big power politics. There is no doubt that reciprocal altruism has been an important feature of our own evolution, and we have ideas of morality and fair play, friendship and gratitude, which are best seen as mechanisms that have evolved to control reciprocal altruism in the most advantageous way for the individual. Since we all evolved under the same evolutionary pressures, and we share the same basic gene pool, it is a reliable rule of thumb in life to apply the tit-for-tat strategy in your interactions with other people: cooperate the first time you interact, and then follow your 'partner's' example for future 'moves' in the 'game'. In other words, 'do as you would be done by'. The result is that most people, most of the time, cooperate with each other and get along fine.

So what can this tell us about superpower politics? Politicians are just people, and the same rules apply to them as to other people. Even if we personify each superpower as one player in a supergame of Prisoner's Dilemma, the rules are still the same. The correct strategy in the arms race scenario is to offer to cooperate with the other guy in reducing arms, and then follow his examples on all subsequent moves. It is now easier than ever to see how we got ourselves into the Cold War mess, and why it proved so easy, in the end, to get out of it. In the unusual circumstances following the Second World War, neither of the new power blocs trusted the other, even though they had just been working together to defeat a common enemy. The mere fact of the existence of new superweapons was enough to induce paranoia, and the 'game' went wrong at the first step. For decades, each side faithfully – and correctly in terms of the strategy – followed the tit-for-tat rule on the up escalator. But once someone had the initiative to take one of the olive branches offered by the other side at face value, it proved just as easy to reverse the process, and proceed on the down escalator by a series of cautious, graded, tit-for-tat *dis*armaments. 'Offer to cooperate, then follow the other guy's lead' is the best rule.

It is fortunate, indeed, that the superpowers seem to have learned this lesson at the beginning of the 1990s. For just as the Cold War has thawed, humankind has been faced with a problem that will demand the cooperation of just about everybody on the planet if it is to be overcome without massive disruption to the society that the 'third chimpanzee' has built up since the end of the last ice age. The biggest threat facing humankind now is not nuclear war, but global warming: a climatic change of our own making, which is now well on the way to bringing back the warmth that the dinosaurs thrived in.

CHAPTER TEN

Dinosaur Days are Here Again

Climates of the past have influenced the evolution of the upright ape, and affected the way in which human societies have spread around the globe. What does the climate of the future hold in store? Climate is always varying, on many different timescales. We can be certain that, if we take a long enough perspective, then further ice ages as well as periods of warmth comparable to the heyday of the dinosaurs lie ahead.

If the ice-age rhythms of the past million years persist, then the next ice age is just about due: next century, or next millennium, if not actually tomorrow. Indeed, if the little ice age that froze the Vikings in Greenland, forced the Amerindians to give up farming, and froze the milk in Shakespeare's milkmaid's pail had intensified into a full ice age, the pattern would have fitted very neatly into the astronomical cycles of ice ages. Perhaps the 'next' ice age should have started yesterday. Will the warming that has pulled us out of the little ice age persist? Or is another tightening of the climatic screw to be expected? Which way is the climate likely to jump on the timescale that matters to us, and our children: over the next hundred years or so?

One way to make an educated guess at what might lie in store is to take the actuarial approach, used by insurance companies and by planners responsible for sea defences, reservoirs and other major works of engineering that are designed to mesh in with the patterns of the weather. We can look back at the recent past to see what extremes of climate have occurred over a reasonable number of centuries, and guess that nothing more extreme will happen over the next hundred years. If we are interested in the next century, then it is a reasonable rule of thumb to look back over the past ten centuries – the past millennium – and guess that there will be no patterns of weather in the twenty-first century that have not already occurred at some time between AD 1000 and the present. Other things being equal (assuming that no major shift of climate is under way), that should be a fair guide. But there is always the proviso that major shifts in climate do occur, like the onset of colder weather that brought a halt to the Viking voyages. As we shall see, there are good reasons to believe that a climate jump is in store, and may already have begun. But this change is not part of any of the natural patterns of climatic variation that we have discussed so far, and in order to put it in its proper perspective it is best to look at how the climate of the twenty-first century might have developed, had people not been around to throw a spanner in the works of the weather machine.

Cycles that follow the Sun

The most obvious feature of the weather of the past millennium, compared with today, is that by and large it was colder. In any plot of temperature trends since AD 1000, the twentieth century stands out as a period of sustained warmth. The difference is small: if we look at averages decade by decade, to smooth out the natural year-to-year variability of weather, all the fluctuations in average temperature over the past few hundred years cover a range of only about 1°C. But those small fluctuations were still significant, as our description of historical events of the past millennium should have made clear. Most people who live in temperate latitudes think that a warmer world is a more pleasant place to live (which is why warm periods of the past get names like 'little optimum'), and on that basis an optimist might hope that the warmth of the twentieth century marks a break-out from the cold of the little ice age, and will be sustained. But the actuarial approach to climate forecasting would suggest just the opposite: that we have just lived through a half-century or so of unusual warmth, and that statistically speaking we are due for another run of cold decades, returning the weather to the normal pattern of the past millennium. Studies of the way in which the climate has changed in the past, and observations of the way the Sun itself has varied from decade to decade and century to century, lend credence to this more gloomy prognosis.

The 'little ice age' is a term that means different things to different people, because the cold of the past few hundred years varied in intensity over the centuries. The most naive view of those patterns of intense cold itself suggests that the warmth of the twentieth century may not last. The little ice age was at its peak intensity in the seventeenth century; then, after a run of decades when the weather was slightly less harsh, it returned in something like its former strength in Dickens' time, in the first half of the nineteenth century. There is a hint in the rhythms of the little ice age of a succession of particularly cold intervals, separated by about 180 to 200 years. On that naive picture alone, the next cold blast ought to be due any time now, since we are already 180 years on from Dickens' boyhood days.

But why should the climate vary in such a regular fashion? What could be producing cyclic vairations in the temperature of the globe? The obvious answer is that something happens to the Sun itself, changing the amount of heat which is available to keep our planet warm. At least, the answer seems obvious to some climatologists; many astronomers have long poured scorn on the notion that the Sun might vary, even by the 1 per cent or so needed to explain these climatic rhythms, on a timescale of decades and centuries. However, evidence has recently been mounting which suggests that the climatologists were right and the astronomers were wrong.

Two hundred years after the French revolution, the Royal Society and the Academie des Sciences held their first joint meeting, to discuss the Earth's climate and the variability of the Sun. They investigated the way in which solar variations, seen most visibly in the varying number of dark spots on the surface of the Sun, affect our weather. And they found that there is clear evidence of a roughly 200-year cycle of solar activity and climate.

The Sun's activity is measured in terms of sunspot number. Dark spots on the surface of the Sun are clearly visible with the aid of a telescope (used to project the image of the Sun onto a white surface) and have been monitored since Galileo first turned a telescope heavenwards in the early seventeenth century. For most of the past 300 years, the spots have come and gone in a fairly regular cycle, roughly 11 years long, known as the sunspot cycle. Twentieth-century observations, including measurements made by instruments on board satellites, show that when there are more spots on its surface the Sun is more active overall, producing flares and streamers of material that break out into space and create a 'solar wind' of particles that streams out past the planets of the Solar System. There are also solar magnetic changes linked with the sunspot cycle. But curiously, there were very few spots visible at all during the peak years of the little ice age, in the second half of the seventeenth century.

Partly because of this, the 11-year cycle has only been recorded accurately by astronomers since about 1700. But much longer records of the varying activity of the Sun are stored in the wood of trees, in the form of radioactive atoms of carbon-14. Carbon-14 is produced in the atmosphere by the interaction of particles from deep space, known as cosmic rays, with atoms of nitrogen-14. Some of these radioactive carbon atoms become part of carbon dioxide molecules and are taken up by trees during photosynthesis and stored in the cellulose of their rings. Because the age of these rings can be determined precisely by counting back through the years, and the rate at which radioactive carbon-14 'decays' into other, stable atoms is known, tree rings provide a record of how much carbon-14 was being produced each year in the atmosphere. With the aid of long-lived trees and overlapping segments of preserved dead wood, dendrochronologists have built up a complete tree-ring and carbon-14 record going back 9000 years, ample to detect variations in carbon-14 production linked with changes in solar activity on timescales of decades and centuries.

The Sun influences carbon-14 production by altering the flow of cosmic rays into the atmosphere of the Earth. When the Sun is *more* active, the solar wind of particles streaming out through the Solar System holds back the cosmic rays, so there is *less* carbon-14 production. The relationship can be checked by comparing the tree-ring record of the past two centuries with direct observations of sunspots. This confirms that the wiggles in the carbon-14 record mirror the changes in solar activity.

Charles Sonett, of the Univeristy of Arizona, Tucson, told the meeting (held at the Royal Society on 15–16 February 1989) that this long carbon-14 record is dominated by a cycle about 200 years long. There are also variations corresponding to rhythms 2300 years long and just under 1000 years long, together with cycles of 80–90 years and the familiar 11-year sunspot cycle, both of which show up in analysis of the astronomical records of sunspot activity.

The special interest in this discovery for climatologists is that the times during this dominant 200-year cycle when the Sun was relatively quiet seem to correspond with decades of cold on Earth, at least over the past millennium, and also, although the evidence is less complete, before AD 1000.

The same 200-year period also shows up in analyses of the widths of tree rings. At intervals of 200 years there is a tendency for the rings to be much narrower for several decades, showing that the trees were experiencing some form of stress. This pattern recurs in the decades just before each peak of the 200-year carbon-14 cycle. All of this evidence strongly suggests that on these occasions a change in solar activity first cooled the world into a little ice age, and then led to a build-up of carbon-14 in the atmosphere as cosmic rays penetrated it more easily.

In isolation, this evidence might be taken with a pinch of salt. But quite separate lines of research point to the same conclusion. Zhentao Xu, of the Purple Mountain Observatory in Nanjing, described an analysis of ancient historical records of solar phenomena (such as sunspots) recorded in China, which suggests that there has been a long-term 210-year cycle of activity over the past 2000 years. And researchers from the Centre de Spectrométrie Nucléaire et de Spectrométrie de Masse in Orsay have found that traces of the long-lived radioactive isotope beryllium-10, found in ice cores drilled from Antarctica, show a cyclic variation with a 194-year period. Since beryllium-10 is also produced in the atmosphere by cosmic rays, but settles to the surface of the Earth without being taken up by living organisms, this provides a separate direct measure of changing solar activity.

Taking all that at face value, it ought to be possible to predict future levels of solar activity, and perhaps short-term temperature trends, by looking at the records from 200 and 400 years ago. At the end of the sixteenth century the world was sliding into the worst decades of the little ice age, and the Sun was in decline. In the 1790s the Sun reached a high peak of activity, but this was followed by three cycles of very low activity, and a temporary return of little-ice-age conditions in the first three decades of the nineteenth century. The recent peak in solar activity has given astronomers and climatologists an ideal opportunity to test the 200-year cycle, checking to see if the pattern of the early nineteenth century repeats in the early twenty-first century, with the Sun becoming quieter and the world becoming cooler.

Unfortunately (not just for them, but for many other people around the globe), the world may continue to warm up, whatever the Sun does over the next half-century. The 200-year solar rhythm can produce only a minor climatic ripple on the surface of longer and larger variations, such as the shift from the post-glacial warmth to the cold of the little ice age (or even the shift from ice age to interglacial). And the warming we are now experiencing is likely to overwhelm any cooling influence linked to a decline in solar activity over the next 50 years. The effect of human activities on climate is already producing as big a fluctuation as any natural change of the past millennium, and the effect is getting bigger all the time.

Into the greenhouse

The 1980s stand out as warmest decade on record, and global temperatures have risen by about 0.5°C since the beginning of the twentieth century. The trend of temperatures over the past 100 years has not been uniformly upward. The world as a whole cooled slightly from 1850 to 1880, then warmed up until the 1940s. There was a slight cooling into the 1970s, but for more than ten years now the warming has returned, stronger than before. In 1988, the world was 0.34°C warmer than the average for the period from 1950 to 1980, while 1987 was 0.33°C warmer than the 1950–80 average. There is clear evidence that although other factors, such as volcanic dust, or changes in the Sun's output, have caused temporary setbacks, the underlying temperature trend is upward.

The global warming also shows up in rising sea levels around the world. Once again, there are variations from year to year, but the long-term trend is clear. Since 1900, the sea level has risen at a rate of 10 to 20 cm per century. This is partly because small mountain glaciers are in retreat, melting to provide more water, which ultimately runs down to the sea. (The great polar ice caps are *not* yet in retreat because when the world warms a little more moisture evaporates from the ocean and is available to fall as snow in the polar regions.) It is also simply because the sea water has expanded as the world has got warmer.

We do not have to look far to find the reason for this warming: it is the profligate way in which we have been burning fossil fuel, ever since the Industrial Revolution in Europe. Coal and oil burnt to provide power for industry, to heat our homes or to drive our cars is mainly carbon, and when it burns it puts carbon dioxide into the air. Human activities have increased the amount of carbon dioxide in the atmosphere by a quarter since the middle of the nineteenth century, and carbon dioxide is a 'greenhouse' gas, one that traps heat from the ground that would otherwise be radiated away into space. The more carbon dioxide there is in the atmosphere, the warmer the world will be, and calculations of the warming effect of the actual increase in carbon dioxide over the past century closely match measurement of the actual increase in temperatures worldwide.

So far, the effect has been small. But the build-up of carbon dioxide is now increasing very rapidly (it is one of those exponential curves that starts off rising slowly, then turns a corner and shoots upward; we are just on the corner today). Many other gases being released by human activities are also greenhouse gases, including CFCs that are also responsible for the destruction of the ozone layer, and methane that is released in large quantities when tropical forests are burned and from the paddy fields that provide the rice that is the staple foodstuff for a large fraction of the world's population. Taking all these gases together, the equivalent of doubling the amount of carbon dioxide present in the atmosphere in 1850 will occur by about 2030 unless drastic measures are taken to halt the build-up in the very near future.

On a timescale of thousands or millions of years, the Earth can adjust to this disturbance. The presence of more carbon dioxide in the air will, for example, increase the rate at which carbonates are deposited, and in time a new equilibrium will be established, as it has been after natural upheavals in the past.

But on a timescale that matters to people, a few tens of years, there is no time for these long, slow processes of adjustment to come into play. The build-up of greenhouse gases projected for the year 2030 implies that global mean temperatures will increase by about 4°C, with a much bigger increase at higher latitudes (nearer the poles) than at the equator. This is a much larger change than any natural climatic shift that has occurred on a similar timescale. Shift into and out of a little ice age requires temperature fluctuations of just a degree, so even if another little ice age is due it will be overwhelmed by the warming trend. A warming by 4°C is on the same scale as the transition from a full ice age to an interglacial, and that takes thousands of years to complete.

Such a warming will change the world dramatically. Climatic zones will shift. There will be more moisture in the air because higher temperatures will increase evaporation from the ocean, so coastal regions in some parts of the world may become wetter, with increased rainfall. At the same time the sea level will be rising, not by 10 cm a century, but by 10 cm a *decade*, inundating low-lying regions such as the Nile Delta, Bangladesh and Florida. But as the temperature rises the continental heartlands will dry out, and deserts will spread in many parts of the world. The Sahel region of Africa, and Ethiopia, may already have felt the bite of the new climatic pattern; southern California and the European Mediterranean may be next in line.

The speed and size of these changes are so great that they give the lie to our cosy notion that a warmer world would be a better place to live. Compared with what is in store for us over the next half-century, a return to the little ice age might be more welcome. The Mediterranean region, from Spain to southern Italy and Greece, will become increasingly arid and desert-like over the next few decades, the extreme heat of the summers of 1987 and 1988 in that part of the world being merely a taste of things to come. Similar problems of desertification will apply to much of the southern USA and Mexico, while the American Great Plains will become too hot and dry to grow corn, even with the aid of modern farming techniques. (The reason for the present global warming, the greenhouse effect, is quite different from the reasons, whatever they were, for the little optimum, so there is no likelihood that the rains which made life pleasant for the Mill Creek people will return.) The flow of the Colorado River is likely to be halved over the next 50 years, according to current greenhouse projections.

But more warmth, and more evaporation from the sea, also mean more storm activity. Storms are driven by the effect of moisture in the air condensing and giving up energy (its latent heat) to drive the wind machine. Hurricanes will become more common, and more severe, as we move into the twenty-first century. So will storms at high latitudes. But some regions of the globe may benefit from the changes that are likely to occur. The US corn and wheat belts could shift northwards into Canada, while less severe weather would help to open up new lands in Siberia for agriculture. Other regions, especially the tropics, may notice little change in the weather. When the world warms the effect is felt more strongly at the poles, because that is where most heat is being lost into space and where the greenhouse effect can have a big impact. In the tropics, where the main influence is heat coming in from the Sun, changes will be much smaller. On balance, though, the changes brought about by this new greenhouse effect, a direct result of human activities, seem likely to cause turmoil in the twenty-first century, as many regions struggle to adjust.

One of the chief problems (even for regions like Canada, which might benefit from a doubling of the natural concentration of carbon dioxide) is that the effect does not stop there. The world will not simply get to be 4°C warmer in 2030 and stay that way. In the twenty-first century, humankind will still be releasing greenhouse gases, making the planet warmer still. Only when we stop releasing those gases (which might not be until the coal runs out, maybe 200 years from now) will it be possible for a new equilibrium to become established, with average temperatures higher than today but with a new, settled pattern of climate zones. And achieving that equilibrium will take 100 years or more, from the day when people cease adding to the brew of greenhouse gases in the air.

What would the world be like then? This book is not about the anthropogenic greenhouse effect, and the problems of the twenty-first century are on too fine a timescale to fit neatly into the broad geological perspective we have been concerned with so far. But we can skip past the turmoil of the twenty-first century to look at what the world might be like a few hundred years from now, after the climate had adjusted to the impact of burning all the fossil fuel. The picture we come up with is strikingly familiar.

Dinosaur days

Remember the dinosaurs? They thrived during an era, the Mesozoic, when the world was on the whole warmer than it is today. Various pieces of evidence show that tropical conditions existed all the way to latitudes of 45° (where we find Minneapolis, Ottawa, Turin and Vladivostok today in the north; all of Australia lies equatorward of this line in the south). Dinosaurs themselves lived not only in those extended tropical regions, but even closer to the poles. But the fact that tropical swamps and river deltas filled with lush vegetation existed in some parts of the world up to latitude 45° does not necessarily mean that every square metre of the land surface of the globe between 45°N and 45°S was covered with tropical swamp during the Mesozoic. Robert Bakker, as provocative on this as in his argument (now widely accepted) that dinosaurs were hot-blooded, has gathered together a wealth of evidence which is consistent with a picture of dry continental interiors where creatures like the brontosaurus roamed, and the lush vegetation and swampy tropical jungle was restricted to low-lying coastal regions, especially river deltas. His picture of the Jurassic world, the period in the middle of the Mesozoic, consists of 'a system of broad, flat floodplains, small rivers, shallow ponds, and occasional deep lakes, all subjected to cycles of killing droughts'.* The big dinosaurs, he says, avoided swamps; and he has a much more satisfactory explanation for the structure of their feet than those based on the old idea of somnolent giants paddling in the mud.

One of the classic pieces of 'evidence' that dinosaurs enjoyed a semi-aquatic lifestyle comes from the fossil remains of a type of dinosaur known as a duckbill, more than three times as long as a man. This fossil is so perfectly preserved that the stony

* Robert Bakker, *The Dinosaur Heresies*, p. 118.

remains still show traces of webs of skin between the creature's toes. What could be more eloquent proof of an aquatic lifestyle than webbed feet? But Bakker has another explanation for the 'webbing'. Camels, he points out, have thick pads like cushions under their toes, to provide support for their feet and act as shock absorbers as they walk or run across dry, hard land surfaces. On a visit to a South African park, Bakker came across the desiccated corpse of a camel, dried out and mummified in the heat. What had once been cushions under the toes of the living camel have become flattened, dried out bags of skin on the mummy, a perfect imitation of a webbed foot. And that, he says, is why the duckbill fossil seems to have webbed feet: what had once been fat shock absorbers under its feet had taken on the appearance of webbed toes after the creature died and its remains mummified in some hot, dry region.

There is, in fact, a great deal more evidence that creatures like the duckbill (and brontosaurus) were adapted to running over dry plains, not to splashing about in muddy swamps. You can find it summarized in Bakker's book, if you want the details. What matters for our own story is the overall picture he paints, of dry continental interiors with swampy coastal margins and lush river deltas. The swamps and river deltas were, of course, the regions in which carbon dioxide was being taken out of the air in large quantities by living plants, and converted into the carbon-rich material of their leaves, stems and branches. Plant remains, buried in those swamps and squeezed and heated as a result of geological activity over millions of years, became rich deposits of coal, some of which are being mined and burnt today, returning the carbon to the atmosphere as carbon dioxide. Could it be possible that the world of the dinosaurs was so warm at least partly because there was more carbon dioxide in the air at that time, strengthening the greenhouse effect?

This speculation has been supported by the work of Keith Rigby, of the University of Notre Dame.* Among other features, studies of the nature of ancient stream beds and the

* He describes the evidence in his contributions to the second volume of *Dinosaurs Past and Present*, edited by Sylvia Czerkas and Everett Olson; the quote is from p. 129.

structure of their banks show that, at least in the region that is now Montana, the dinosaurs of the Cretaceous period lived in an environment with a pronounced dry season, subjected to repeated flash flooding. The evidence confirms that the Cretaceous world was substantially warmer than today, but also shows that 'the dinosaurs were living in a much different environment than the swamps in which we typically see them illustrated'.

Just how hot was the world at this time? According to a variety of evidence summarized by Rigby, North America may have been as much as 25°C warmer than it is today, while the equatorial region was only about 5°C warmer. The pattern is exactly the same as the pattern of greenhouse warming seen in computer projections of how the weather is likely to change in the twenty-first century, with higher latitudes warming the most. And that pattern may indeed have persisted in the days of the dinosaurs for the same reason. We mentioned before how the shifting positions of the continents has altered the climate of the globe over geological time, moving landmasses towards or away from the poles, diverting the flow of ocean currents, and either allowing warm water to keep the polar regions ice-free or blocking off the flow of warm water and causing the poles to freeze. But this effect can account for only part of the difference between the present-day climate and the warmth of the Cretaceous. The geography of the globe was not so very different then from the geography today, and experts estimate that the difference could account for a warming of North America of no more than 5°C. The other 20°C of warming is best explained as a result of a greenhouse effect.

That in itself takes us back to continental drift and geological activity. The extra carbon dioxide in the air in the day of the dinosaurs must have come originally from volcanoes and from outgassing at cracks in the Earth's surface. Climatic reconstructions suggest that in the Cretaceous there may have been as much as ten times more carbon dioxide in the air than there is today. Carbon dioxide, as well as being a greenhouse gas, is an essential ingredient for plant growth, one of the main ingredients (along with sunlight and water) in the process of photosynthesis. Researchers who study the implications of the present build-up of carbon dioxide in the atmosphere caused by human activities talk of a 'carbon-dioxide fertilization effect', and suggest that many plants will grow more vigorously as the carbon-dioxide concentration increases. (Unfortunately, this may not be an unmixed blessing for farmers; in many cases, it seems that weeds will benefit far more than food crops.) A strong carbon-dioxide fertilization effect operating in the Cretaceous would very neatly explain the increased biological activity of plants that took so much carbon out of the air and locked it up in coal deposits at that time.

The overall picture that Rigby paints is of a world very much warmer than today, in which there was more evaporation from the oceans and thus more rainfall overall, perhaps 25 per cent more rain each year than today, worldwide. But most of that rain would have fallen not on the plains but on the coastal areas, which became swampy, tropical jungles while the continental interiors became dry and subject to flash flooding, exactly in line with Bakker's picture. The last of the dinosaurs, says Rigby, lived in an environment very similar to that along the Fitzroy River of Western Australia today. Dense, tropical vegetation lines the course of the river itself; but away from the river the land is dry, and the river itself dries up at some times of year. If this interpretation is correct, then this region of western Australia provides the best guide to the kind of climate North America and Europe can expect as the anthropogenic greenhouse effect takes a grip on the world. But the story also provides some new insights into how and why the dinosaurs became extinct.

According to the palaeoclimatic evidence, the average temperature of the globe was about 25°C around 100 million years ago. It declined fairly steadily to about 20°C at 80 Myr, and around 17°C at the time of the death of the dinosaurs, at 65 Myr. Over that same span of time the number of genera of dinosaurs declined from more than 30 before 80 Myr to about 15 around the end of the Cretaceous. There is, as we mentioned earlier, a wealth of evidence that environmental changes, especially a cooling of the globe, had sent the dinosaurs into retreat long before the final disaster struck. Although we are impressed by the evidence that there was indeed a final disaster, probably a meteorite impact, that marked the end of the Cretaceous, it seems likely that this event was so effective at writing an end to the day of the dinosaurs because their grip had already been weakened by the progressive cooling of the globe in the Late Cretaceous. Some of that cooling was produced by the rearrangement of the continents. But just as continental drift alone cannot explain why the world was so warm at 100 Myr, so continental drift alone cannot explain why it subsequently cooled so much. If it was indeed the greenhouse effect that made the world of the dinosaurs so warm, as all the evidence suggests, then clearly it must have been a weakening of the greenhouse effect that made it cool. Probably the volcanoes became slightly less active at some stage in the geological story, with less carbon dioxide getting in to the atmosphere each year. But for millions of years the lush tropical vegetation of the coastal regions of the hothouse world would have continued to absorb carbon dioxide from the air and lay it down as coal. By the time this biological activity had reduced the carbon dioxide content of the atmosphere so much that the world was becoming noticeably cooler, the end of the day of the dinosaurs was in sight. A new climatic balance would ultimately be struck, one which would turn out to favour mammals, not dinosaurs.

Our story has come full circle, and this is literally where we came in. We are the children of the ice, the heirs to the dinosaurs, and we owe our existence to the climatic changes that brought about the demise of the dinosaurs. Those climatic changes were closely linked to biological activity, taking carbon dioxide out of the air and storing it as coal, and so weakening the greenhouse effect and cooling the globe. Now, however, human activities are reversing the trend. We are busily taking carbon, in the form of coal, out of the ground, and burning it to make carbon dioxide, which is building up in the atmosphere and warming our planet. We are restoring the kind of climate that suited the dinosaurs so well, conditions in which our ancestors were reduced to scurrying about in the undergrowth, surviving in the shadow of the dinosaurs. Dinosaur days are indeed here again. The whole pattern of climate and evolution over the past 65 million years – the unusual climatic circumstances that brought about our own existence – may be seen, in the long story of the Earth itself, as no more than a temporary aberration caused by too much carbon being taken out of circulation for a time. Although the dinosaurs themselves are no longer around to benefit from the change back to 'normal' weather conditions on Earth, if those hotter conditions persist they could well leave a mark in the geological record as significant as the Cretaceous-Tertiary boundary itself. It is time for us to take our leave, with just a passing curiosity about what palaeontologists 65 million years from now (if there are any) will make of this new blip in the geological record.

BIBLIOGRAPHY

Books mentioned in the text, and other relevant books, are detailed here. Ones marked with an asterisk are less of an easy read than the others, but will repay the extra effort involved in getting to grips with them. Editions referred to are simply the ones that came to hand while we were writing this book; the town of publication is given for those not published in the UK.

R.D. Alexander, *Darwinism and Human Affairs*, University of Washington Press, Seattle, 1979.

Robert Bakker, *The Dinosaur Heresies*, Longman, 1987.
Our favourite dinosaur book. The inside story of Bakker's now not-so-heretical idea that many dinosaurs were hot-blooded.

David Barash, *Sociobiology and Behavior,* second edition, Elsevier, 1982.
Although this is a textbook, it is quite accessible to the lay-person (and does not require an asterisk!) and provides a very clear account of the basic principles of sociobiology. Trivers's book is an even better read, but Barash's is more comprehensive and provides an invaluable detailed bibliography.

Connie Barlow (editor), *From Gaia to Selfish Genes*, MIT Press, Cambridge, MA, 1991.
A very readable collection of articles touching on many of the themes addressed in our book.

Steven Brams, *Game Theory and Politics*, Macmillan, 1975

* Jere Brophy and Sherry Willis, *Human Development & Behaviour*, St Martin's Press, New York, 1981.
A textbook for those who want to delve deeper into

subjects such as the nature/nurture controversy, social development, and so on. Not for the casual reader.

Reid Bryson and Thomas Murray, *Climates of Hunger*, University of Wisconsin Press, Madison, WI, 1977.
A short, readable overview of the impact of climatic change on many human societies, focusing especially on the North Atlantic Viking colonies and the troubles experienced by the Amerindian cultures over the past 1000 years or so. Bryson is one of the world's leading climatologists; Murray is a professional writer. The combination works well.

Nigel Calder, *Timescale*, Chatto & Windus, 1984
The best single-volume chronology of everything important to human evolution, starting with the birth of the Universe in a Big Bang between 10 billion and 15 billion years ago. Calder tends to invoke cosmic impacts to explain just about every extinction in the fossil record, and this enthusiasm for cosmic catastrophe may not be fully justified. But overall the book absolutely justifies its subtitle, 'an atlas of the fourth dimension.'

Arthur Caplan (editor), *The Sociobiology Debate*, Harper & Row, 1978.
A collection of readings from both sides of the debate that raged during the mid-1970s. Now slightly out of date, but a useful historical guide from which you can draw your own conclusions about who has the sounder scientific arguments.

Jeremy Cherfas and John Gribbin, *The Redundant Male*, Pantheon, New York and Grafton, London, 1984.
A detailed look at human sex, making the case that men have outlived their evolutionary 'purpose'.

* Russell Ciochon and Robert Corruccini (editors), *New Interpretations of Ape and Human Ancestry*, Plenum, New York, 1983.
A weighty (888 pages) scientific tome which includes some fascinating and readable chapters on the molecular clock and on the pygmy chimpanzee. Worth digging out of a library.

Preston Cloud, *Oasis in Space*, Norton, New York, 1988.

A history of the Earth, mainly from a geological perspective but with some insight into the evolution of life.

Paul Colinvaux, *Why Big Fierce Animals are Rare*, Pelican, 1980.
An excellent, readable account of how the living world works.

Paul Colinvaux, *The Fates of Nations*, Pelican, 1983.
Slightly less accessible than *Big Fierce Animals* and with a more 'serious' somewhat controversial message about historical imperatives, derived from the author's training in ecology. A thought provoking perspective on humankind as an animal species.

John Crane (editor), *Annual Editions: Biology*, Dushkin, Hartford, CT, 1984.

James Croll, *Climate and Time*, Appleton, New York, 1875.
Early history of the astronomical theory of ice ages, from the pioneer of the subject. Interesting, but not essential reading since Croll's story is summarized very well by John Imbrie and Katherine Imbrie's *Ice Ages: Solving the Mystery*.

* Helena Curtis, *Biology*, second edition, Worth, New York, 1975.
Our favourite overview of biology, because it is the best. No less than 1065 pages, with something about anything you want to know about life, including the topics discussed in our book. There is a more recent edition, so the one we have is therefore a little out of date, but it is an old friend. For those with a serious interest, however.

* Sylvia Czerkas and Evertt Olson (editors), *Dinosaurs Past and Present*, 2 volumes, Natural History Museum of Los Angeles/University of Washington Press, Seattle and London, 1988.
The asterisk refers to the text, and the books are a little pricey for the layperson, but we recommend them chiefly for the astonishing paintings and drawings, scrupulously accurate scientific reconstruction of how dinosaurs must have looked. Dale Russell's 'dinosauroid' features in Volume 1. Bully your local library into buying them.

Martin Daly and Margo Wilson, *Sex, Evolution, and Behaviour*, Duxbury Press, Belmont, CA, 1978.

Perhaps just technical enough to merit half an asterisk, but the most definitive reasonably accessible account of what sex is all about.

Charles Darwin, *The Origin of Species* and *The Descent of Man*, Random House Modern Library, no date given.

Curiously, this modern reprint of both Darwin's classics in a single volume at an incredibly low price ($8.95 when we bought it in 1985) does not carry a publication date. The edition of the *Origin* is the sixth, from 1872, while the version of the *Descent* is the first, from 1871. The whole runs to exactly 1000 pages and includes Darwin's lengthy discussion of selection in relation to sex. He wrote beautifully, with great clarity, and anyone interested in evolution and human origins should find these originals fascinating. The text of the first edition of the *Origin*, which in some ways presents Darwin's ideas most clearly, can be found in a Penguin edition, reprinted most recently in 1993.

Richard Dawkins, *The Extended Phenotype*, Oxford University Press, 1983 (original edition published by Freeman, New York, 1982).

A delicious book in which Dawkins, a superb writer, elaborates on the theme he developed originally in his more 'popular' account, *The Selfish Gene* (Oxford University Press, 1976). Taking on board, and responding to, the criticism that book received, he has produced a book that is of major significance to professional biologists but also entirely intelligible, and enjoyable, for the interested onlooker. It contains probably the best and most convincing evidence in broad support of the ideas of human sociobiology.

Richard Dawkins, *The Blind Watchmaker*, Penguin, 1988.

The best book about evolution for the general reader.

Jared Diamond, *The Rise and Fall of the Third Chimpanzee*, Hutchinson Radius, 1991.

A more up-to-date account of the relationship between humans and chimps, along the same lines as the Gribbin and Cherfas book (see below).

David Dinely, *Earth's Voyage Through Time*, Granada, 1974.
Primarily a history of the changing face of the 'solid' Earth, but with major events in the development of life put in their broad environmental context.

Niles Eldredge, *Life Pulse*, Facts on File, New York, 1987.
A very readable account of bits of the fossil record that interest Eldredge, and that he has studied himself.

Kurt Fischer and Arlyne Lazerson, *Human Development*, Freeman, New York, 1984.
Marketed as a textbook, but completely intelligible to the intelligent layperson, clear and comprehensive. But more something to borrow from a library than one to burden your own bookshelf with permanently: over 700 large-format pages on everything from heredity to the problems of adolescence.

Stephen Jay Gould, *The Mismeasure of Man*, Pelican, London, 1984 (original edition published by Norton, New York, 1981).
The best book we know of about the IQ debate. As good a read as many a work of fiction, with astonishing insights into the way all too many scientists allowed their prejudices to influence their work. Unfortunately, in the final chapter of the book Gould allows his own prejudice against sociobiology to show through and cloud his judgement, but this does not affect his compelling history of the study of human intelligence.

John Gribbin, *In Search of the Double Helix*, Corgi, London and Bantman, New York, 1985.
The story of evolution from Darwin to DNA, and beyond into the modern understanding of genetics and the basis of life.

John Gribbin, *Hothouse Earth*, Bantam, 1990.
The anthropogenic greenhouse effect described in detail.

John Gribbin and Jeremy Cherfas, *The Monkey Puzzle*, Paladin/Granada, 1983.
An account of the development of the 'new', and now established time-scale of human development from the ape line.

Mary Gribbin (with John Gribbin), *Too Hot to Handle?*, Corgi, 1992.

More about the greenhouse effect (see Chapter Ten).

Harry Harrison, *West of Eden*, Panther, 1985.

The first part of a science-fiction trilogy, and based on the idea that the dinosaurs were not wiped out 65 million years ago but continued to evolve and developed intelligence. Entertaining hokum with a smidgeon of evolutionary insight.

Ernest Hilgard, Rita Atkinson and Richard Atkinson, *Introduction to Psychology*, seventh edition, Harcourt Brace Jovanovich, Orlando, FL, 1979.

Known to generations of psychology students as 'Hilgard and Atkinson', this is the undergraduate development psychologist's bible, and by no means inaccessible to the more casual reader who might want to dip into it in a library.

John Imbrie and Katherine Palmer Imbrie, *Ice Ages: Solving the mystery*, Macmillan, 1979.

A satisfying complete and enjoyably intelligible account of the development of the Milankovich model of ice ages, which explains how changes in the Earth's orientation in space produce climatic cycles and ice-age rhythms.

Donald C. Johanson and Maitland A. Edey, *Lucy*, Granada, 1981.

An exciting, highly readable account of the search for fossil remains in Ethiopia and of the discovery and interpretation of the fossils of Lucy, a human ancestor that lived 3 million years ago and looked very much like the present-day pygmy chimp. Johanson's pet theories on the mechanisms of evolution, and particularly the role of sex in human evolution, are, however, decidedly suspect and should be taken with a pinch of salt.

Gwyn Jones, *A History of the Vikings*, Oxford University Press, 1968.

Anything by Gwyn Jones on the Vikings is worth reading. This book provides a wealth of background relevant to the Viking voyages discussed in Chapter Six.

H. H. Lamb, *Climate, History and the Modern World*, Methuen, 1982.

A *slightly* technical but still readable and very complete

account of the influence of climate on human activities since the end of the last ice age.

Richard Leakey, *The Making of Mankind*, Michael Joseph, 1981.
The book of the TV series in which master fossil-hunter Leakey recounted the then-accepted view of human origins.

Richard Leakey and Roger Lewin, *Origins*, Macdonald & Jane's, 1977.
The forerunner of Leakey's *Making of Mankind*, and a rather better book, even though it is slightly older and therefore a little out of date. First class on language and intelligence, and worth reading simply for the chapter 'Aggression, Sex and Human Nature.'

Richard Leakey and Roger Lewin, *Origins Reconsidered: In Search of What Makes Us Human*, Little, Brown, 1992.

Roger Lewin, *Human Evolution*, Basil Blackwell, 1984.
The best 'instant guide' to evolution and human origins, very clearly written in short chapters and with many excellent illustrations. Completely up to date, with the molecular-clock findings included, but no specific mention of *Pan Paniscus*.

Roger Lewin, *Bones of Contentions*, Simon & Schuster, 1987.
A gossipy and entertaining account of the discoveries of fossil remains of our ancestors, and of the controversies surrounding those discoveries and their interpretation.

James Lovelock, *Gaia*, Oxford University Press, 1979; *The Ages of Gaia*, Oxford University Press, 1988.
If you are interested in the evolution of life on Earth, you must read these.

Milutin Milankovich, *Durch ferne Welten und Zeiten*, Koehler and Amalang, Leipzig, 1936.
OK, we really only put this one in to show off and impress anyone who does not read German. You are much better off getting the story of ice-age cycles from the Imbries' book, unless you have a serious interest in the history of science and want it from the horse's mouth.

Elaine Morgan, *The Aquatic Ape*, Souvenir, 1984.

The theory that the selection pressures that turned an African ape into *Homo Sapiens* may have owed much to a period of time spent by our ancestors on the shore of lake or ocean, finding food by diving. Very well written, although Morgan's dates for key events in human evolution are somewhat out of date: she does not take full account of the molecular evidence, although in fact the correct dating by no means invalidates her hypothesis. The argument is now receiving some attention from researchers studying human origins, having originally been dismissed as crazy.

Elaine Morgan, The Descent of Woman, revised edition, Souvenir, 1985.

B. Newmand and M. Dando (editors), *Nuclear Deterrence: Implications and Policy Options for the* 1980s, Castle House, 1982.

* H. G. Owen, *Atlas of Continental Displacement*, 200 Million Years to the Present, Cambridge University Press, 1983.

A gorgeous book which includes maps showing phases in the break-up of Pangea. The text is aimed at academics, but the maps are fascinating in their own right. Worth digging out of a library.

J. M. Roberts, *The Hutchinson History of the World*, Hutchinson, 1976.

The best single-volume history of the world. Although Roberts does not acknowledge the importance of climatic change in helping to shape civilization, it is fascinating to compare the major developments he chronicles with the known pattern of climatic fluctuations, which we outline in this book and which are described in more detail by Hubert Lamb.

Steven Rose, Leon Kamin and R.C. Lewontin, *Not in Our Genes*, Pelican, 1984.

Carl Sagan, *The Dragons of Eden*, Coronet, 1978.

Perhaps Sagan's best book; certainly his most provocative. Subtitled 'Speculations on the evolution of human intelligence', which sums it up.

* John Maynard Smith, *The Evolution of Sex*, Cambridge University Press, 1978.
It seems 1978 was a good year for books about the evolution of sex, but this one is much more mathematically technical, and less readable, than Daly and Wilson's.

* John Maynard Smith, *Evolution and the Theory of Games*, Cambridge University Press, 1982.
The definitive study which establishes the mathematical foundations of the evolutionary stable strategy (ESS) idea. Mostly rather technical.

Sally Springer and Georg Deutsch, *Left Brain, Right Brain*, Freeman, San Francisco, 1981.
We once planned to write a book about left-handedness, but changed our minds when we read this one and realized that we could never hope to improve on it. The best book we know of about the two brains that lie within each human skull, slightly 'academic' in tone but very accessible.

Steven M. Stanley, *Extinction*, Scientific American Library, Freeman, New York, 1987.
Crises for life on Earth, put in the perspective of climatic change and tectonic events. Beautifully illustrated, and authoritative. The best up-to-date account of the series of catastrophes that punctuates the evolutionary record of life on Earth.

G. Ledyard Stebbins, *Darwin to DNA, Molecules to Humanity*, Freeman, San Francisco, 1982.
An up-to-date account of evolutionary ideas; non-technical but thorough.

* Randall Susman (editor), *The Pygmy Chimpanzee*, Plenum, New York, 1984.
The definitive, up-to-date nitty gritty on *Pan paniscus*. But mostly rather heavy-going; not for the faint hearted.

Robert Trivers, *Social Evolution*, Benjamin, New York, 1985.
More of a personal view than Barash's textbook, focusing on topics of special interest to Trivers, but a well-written and fascinating account of sociobiological ideas from a researcher who provided some of the key ideas in the 1970s. The best up-to-date account we know of.

Michael White & John Gribbin, *Darwin: A Life in Science*, Simon & Schuster, London, 1995.

* T. M. L. Wigley, M. J. Ingram and G. Farmer (editors), *Climate And History*, Cambridge University Press, 1981.
Mainly for academics, this is the proceedings of a conference held at the University of East Anglia in 1979. Thomas McGovern's study of the Norse colony in Greenland is especially interesting; the whole book is worth dipping into in a library.

John Noble Wilford, *The Riddle of the Dinosaur*, Knopf, 1985.
A comprehensive overview of the dinosaur story, from the first discovery of dinosaur fossils to the controversies about disaster from space.

Edward O. Wilson, *Sociobiology*, abridged edition, Harvard University Press, Cambridge, MA, 1980.
Even the abridged edition runs to 366 large-format pages of sometimes highly technical argument, and only the last chapter specifically discusses human sociobiology. But, like Darwin's books, anyone interested in our own origins will find this book enthralling, as well as essential reading.

Edward O. Wilson, *On Human Nature*, Harward University Press, Cambridge, MA, 1978.
A much more accessible account of the application of sociobiology to people.

V. C. Wynne-Edwards, *Animal Dispersion in Relation to Social Behaviour*, Hafner, New York, 1972.

Robert M. Yerkes, *Almost Human*, Cape, 1925.

INDEX